Place as Occupational Histories

An investigation of the deflated surface archaeological record of Pine Point and Langwell Stations, Western New South Wales, Australia

Justin Shiner

BAR International Series 1763
2008

Published in 2019 by
BAR Publishing, Oxford

BAR International Series 1763

Place as Occupational Histories

ISBN 9781407302522 paperback
ISBN 9781407332529 e-book

DOI https://doi.org/10.30861/9781407302522

A catalogue record for this book is available from the British Library

This book is available at www.barpublishing.com

BAR Publishing is the trading name of British Archaeological Reports (Oxford) Ltd. British Archaeological Reports was first incorporated in 1974 to publish the BAR Series, International and British. In 1992 Hadrian Books Ltd became part of the BAR group. This volume was originally published by Archaeopress in conjunction with British Archaeological Reports (Oxford) Ltd / Hadrian Books Ltd, the Series principal publisher, in 2008. This present volume is published by BAR Publishing, 2019.

BAR
PUBLISHING

BAR titles are available from:

BAR Publishing
122 Banbury Rd, Oxford, OX2 7BP, UK
EMAIL info@barpublishing.com
PHONE +44 (0)1865 310431
FAX +44 (0)1865 316916
www.barpublishing.com

Table of Contents

List of Figures and Tables iv

Abstract ix

Acknowledgements x

Chapter One: Introduction to the Study 1

Chapter Two: Time and the Temporality of Artefact Accumulation 13

Chapter Three: Studying Stone Artefact Assemblage Composition 23

Chapter Four: Environmental Context of the Study Area 37

Chapter Five: Archaeological Survey and Recording Methods 49

Chapter Six: Establishing a Chronological Context for the Pine Point-Langwell Surface Archaeological Record 59

Chapter Seven: Raw Material Representation and Technological Description 71

Chapter Eight: Raw Material Reduction and Assemblage Composition 81

Chapter Nine: The Temporal Character of Assemblage Formation 107

Chapter Ten: Conclusion to the Study 121

References 129

List of Figures and Tables

Figures

Figure 1.1 Location of the Pine Point-Langwell study area and other features of Western New South Wales mentioned in the text, including the location of assemblages from Fowlers Gap (ND and SC) and Burkes Cave. 3

Figure 4.1 Land systems of the Pine Point-Langwell study area. 43

Figure 5.1 Location of the sampling areas and the SQ1 silcrete quarry. 54

Figure 6.1 Calibrated radiocarbon determination plot for the Pine Point-Langwell hearths. Likelihood is represented by the solid black bars and the posterior by white fill. 64

Figure 6.2 Distribution of dated hearths from the Pine Creek and Rantyga Creek alluvial systems. 65

Figure 6.3 Calibrated radiocarbon determinations from hearths on Fowlers Gap Station. The column on the left contains the hearth code and the un-calibrated determination (Information provided by Dr. Simon Holdaway and Dr. Patricia Fanning). ... 68

Figure 7.1 Distribution of potential raw material sources relative to the four sampling areas. 72

Figure 8.1 Shape plot of clast complete flakes by assemblage. This indicates that clast flakes from the CN3 assemblage on average have a larger surface area to thickness ratio clast silcrete flakes from the other assemblages. 97

Figure 8.2 Shape plot of non-clast complete flakes by assemblage. The small number of flakes in some assemblages creates a variable pattern. However, the plot indicates that non-clast flakes from the KZ1 assemblage on average have a larger surface area to thickness ratio than non-clast flakes from the other assemblages. 98

Figure 8.3 Shape plot of quartz complete flakes by assemblage. The plot indicates that quartz flakes from the CN3 assemblage on average have a larger surface area to thickness ratio than quartz flakes from the other assemblages. However, flakes in the four assemblages are very similar in shape and it is difficult to clearly separate them. 98

Tables

Table 1.1 Radiocarbon determinations from the four rockshelters in Veth's Rudall River study (Adapted from Veth 1993: Chapter Five). The wide range of determinations between the rockshelters indicates that basal occupation was not contemporaneous. Both Karlamilyi and Jalpiyari have evidence of use far earlier than Winakurijuna and Yulpul, therefore as a group the rockshelters do not form a single settlement system. 8

Table 3.1 Water permanency class per individual site and the proportion, and distribution of local raw materials reported by Veth (1993:29-38). Variability in the proportion of local raw materials from individual locations within the same water permanency class indicate that there is no clear relationship between increased amounts of non-local materials at locations with greater water permanency. 31

Table 4.1 Description of the geological history of the Broken Hill Block (Adapted from Stevens 1986:75-78). 40

Table 4.2 Monthly rainfall total (mm) for Pine Creek station from 1991 to 2002. These indicate considerable variability in the amount of rainfall received each month and each year. The total average rainfall over this period is 226.4±124.9 millimetres. 44

Table 4.3 Potential chronological contexts of landforms in the Pine Point-Langwell study area. 46

Table 5.1 Archaeological assessment of individual landforms within each land system for the Pine Point-Langwell study area. .. 51

Table 5.2. Transect CN1A description 52

Table 5.3 Transect CN1B description 52

Table 5.4 Transect KZ1 description 52

Table 5.5 KZ2 Transect description 53

Table 5.6 Description of artefact categories used for the initial quantification of the composition of the Pine Point-Langwell assemblages 58

Table 6.1 Number and percentage (in brackets) of hearth condition in the Pine Creek and Rantyga Creek alluvial systems. The majority of hearths in both alluvial systems were classified as eroding. 60

Table 6.2 Number and percentage (in brackets) of hearth excavation potential recorded in the Pine Creek and Rantyga Creek alluvial systems. The majority of hearths in both alluvial systems were classified as having a high excavation potential. 60

Table 6.3 Radiocarbon determinations on charcoal from heat retainer hearths on Pine Point and Langwell Stations. Weight refers to cleaned charcoal. 63

Table 6.4 Calibrated determinations at the 95% probability level for the Pine Point-Langwell study area. 63

Table 6.5 Interpretation of Bayes factor Model 2 *v.* Model 1, from Rafterty (1996). 66

Table 6.6 Single versus multiple phases of hearth construction tested for the Pine Point-Langwell determinations. Phase A represents the null hypothesis of a single phase, Phase B (five), Phase C (three) and Phase D (two). 66

Table 6.7 Mean likelihood ratio for Single versus multiple phases of hearth construction tested for the Pine Point-Langwell determinations. 66

Table 7.1 Raw material proportion and percentage (in brackets) calculated by number, MNF and volume for the SQ1 assemblage. 73

Table 7.2 Frequency and percentage (in brackets) of clast silcrete non-cortical and cortical complete flakes and cores at SQ1. This indicates that cortex was recorded on the majority of clast silcrete cores and a large proportion of clast silcrete complete flakes in the quarry assemblage. 73

Table 7.3 Raw material by minimum number of flakes (MNF) for the four assemblages. 75

Table 7.4 Raw material by number of individual pieces for the four assemblages. 75

Table 7.5 Raw material by volume (mm^2) for the four assemblages. 75

Table 7.6 Non-gibber to gibber cortex ratio for complete flakes, complete tools and cores combined. A ratio less than one indicate the exploitation of predominantly outcrop sources. Clast silcrete was primarily obtained from outcrop sources, while non-clast silcrete at CN1 and CN3 was mostly obtained from gibber sources. Quartz was obtained from gibber sources for all the assemblages. 76

Table 7.7 Frequency and percentage (in brackets) of clast silcrete complete flakes with different amounts of cortex. The proportion of cortical complete flakes is low in each of the assemblages and flakes with low proportions of cortex dominate this. These results suggest that a significant amount of nodule decortification occurred elsewhere. 77

Table 7.8 Frequency and percentage (in brackets) of non-clast silcrete complete flakes with different amounts of cortex. The proportion of cortical complete flakes is low in each of the assemblages and flakes with low proportions of cortex dominate this. These results suggest that a significant amount of nodule decortification occurred elsewhere. 77

Table 7.9 Frequency and percentage (in brackets) of clast silcrete complete flakes from the SQ1 silcrete quarry with different amounts of cortex. This assemblage exhibits a long-term pattern of raw material exploitation different to that of the four creek assemblages. Cortical flakes account for a much larger proportion of flakes at SQ1, indicating on source reduction. 77

Table 7.10 Frequency and percentage (in brackets) of quartz complete flakes with different amounts of cortex. Quartz exhibits a different pattern to both the silcretes. Cortical complete flakes are common and a large proportion has greater than 50 percent cortex. This indicates on site reduction of cortical nodules. 77

Table 7.11 Length of complete non-cortical and cortical complete flakes from the CN1 assemblage. Clast silcrete cortical complete flakes have the largest length followed by quartz and non-clast silcrete. For each of the raw materials cortical flakes are on average longer than non-cortical flakes. Clast t=1.86, *df*=464, p=0.063, non-clast t= -1.92, *df*=14.04, p=0.076, quartz t=8.52, *df*=2709.98, p=<0.001. 78

Table 7.12 Length of complete non-cortical and cortical complete flakes from the CN3 assemblage. Clast silcrete cortical complete flakes have the largest length followed by quartz and non-clast silcrete. For clast silcrete and quartz cortical flakes are on average longer than non-cortical flakes, however this is not true of non-clast silcrete. Clast t=1.73, *df*=22.67, p=0.097, non-clast t=-0.537, *df*=71, p=0.593, quartz t=7.01, *df*=1621.25, p=<0.001. 78

Table 7.13 Length of complete non-cortical and cortical complete flakes from the KZ1 assemblage. Clast silcrete cortical complete flakes have the largest length followed by quartz and non-clast silcrete. For each of the raw materials cortical flakes are on average longer than non-cortical flakes. Clast t=2.58, *df*=44.15, p=0.13, non-clast t=.263, *df*=13.37, p=0.796, quartz t=3.81, *df*=541.33, p=<0.001. 79

Table 7.14 Length of complete non-cortical and cortical complete flakes from the KZ2 assemblage. Clast silcrete cortical complete flakes have the largest length followed by quartz and non-clast silcrete. For each of the raw materials cortical flakes are on average longer than non-cortical flakes. Clast t=6.07, *df*=180.04, p=<0.001, non-clast t=.822, *df*=212, p=0.412, quartz t=7.81, *df*=2318.58, p=<0.001. 79

Table 7.15 Results of least significance difference test (LSD) for complete cortical flake length between the three major raw material categories within the four assemblages. Only those results that returned a significant difference are displayed. 79

Table 7.16 Results of least significance difference test (LSD) for complete cortical flake length between within raw material class between assemblages. Only those results that returned a significant difference are displayed. 79

Table 8.1 Frequency and proportion of clast silcrete core types per assemblage. Unifacial cores are the most common form. The combined proportion of rotated forms (bifacial and multiple) is highest at KZ1. 83

Table 8.2 Frequency and proportion of non-clast silcrete core types per assemblage. Unifacial cores are the most common form in all assemblages except KZ1, where the proportion of bifacial forms is greater. Again the combined proportion of rotated forms is highest at KZ1. 83

Table 8.3 Frequency and proportion of quartz core types per assemblage. Unifacial cores are the most common form, and the proportion of rotated forms is again highest at KZ1. 84

Table 8.4 MNF (minimum number of flakes) to core ratio per raw material and assemblage. The value for quartz is lowest in each of the assemblages, while the value for non-clast silcrete is highest in each assemblage except KZ1 indicating that there are proportionally more flakes per core for non-clast silcrete compared to the other raw materials. 85

Table 8.5 Non-cortical to cortical complete flake ratio per raw material and assemblage. Quartz has the lowest ratio for each of the assemblages and at CN1 and CN3 there are proportionally more cortical flakes than non-cortical flakes. Clast silcrete has the highest ratio, a result that does not follow that of the MNF to core ratio. 85

Table 8.6 Non-cortical to cortical core ratio per raw material and assemblage. Quartz has the lowest ratio in each assemblages (except non-clast silcrete at CN3) indicating that there are more cortical than non-cortical cores. Again, proportionally more clast silcrete cores do not contain cortex than cores on the other two raw materials. 85

Table 8.7 Mean maximum length, width and thickness of clast silcrete cores in the four assemblages. Maximum length F=0.68, *df* 3, 197, p=0.564, maximum width F=1.87, *df* 3, 197, p=0.136, maximum thickness F=2.05, *df* 3, 197, p=0.108. 86

Table 8.8 Mean maximum length, width and thickness of non-clast silcrete cores in the four assemblages. Maximum length F=2.20, *df* 3, 45, p=0.101, maximum width F=2.18, *df* 3, 45, p=0.103, maximum thickness F=2.25, *df* 3, 45, p=0.096. 87

Table 8.9 Mean maximum length, width and thickness of quartz cores in the four assemblages. Maximum length F=9.33, *df* 3, 1318, p=<0.001, maximum width F=9.50, *df* 3, 1318, p=<0.001, maximum thickness F=10.99, *df* 3, 1318, p=<0.001. 87

Table 8.10 Results of least significance difference test (LSD) for cores on different raw materials in the CN1 assemblage. Only those results that returned a significant difference are displayed. 88

Table 8.11 Results of least significance difference test (LSD) for cores on different raw materials in the KZ1 assemblage. Only those results that returned a significant difference are displayed. 88

Table 8.12 Results of least significance difference test (LSD) for cores on different raw materials in the KZ2 assemblage. Only those results that returned a significant difference are displayed. 88

Table 8.13 Results of least significance difference test (LSD) for clast silcrete cores between assemblages. Only those results that returned a significant difference are displayed. 89

Table 8.14 Results of least significance difference test (LSD) for non-clast silcrete cores between assemblages. Only those results that returned a significant difference are displayed. 89

Table 8.15 Results of least significance difference test (LSD) for quartz cores between assemblages. Only those results that returned a significant difference are displayed. 89

Table 8.16 Frequency and percentage (in brackets) of exterior platform surface of complete flakes per raw material for the CN1 assemblage. Flake scars are the most common exterior platform surface for each raw material. Trimming is most common on non-clast silcrete, followed by clast. 90

Table 8.17 Frequency and percentage (in brackets) of exterior platform surface of complete flakes per raw material for the CN3 assemblage. Flake scars are the most common exterior platform surface for each raw material. Trimming is most common on non-clast silcrete, followed by clast. 90

Table 8.18. Frequency and percentage (in brackets) of exterior platform surface of complete flakes per raw material for the KZ1 assemblage. Flake scars are the most common exterior platform surface for each raw material. Trimming is most common on non-clast silcrete, followed by clast. 90

Table 8.19. Frequency and percentage (in brackets) of exterior platform surface of complete flakes per raw material for the KZ2 assemblage. Flake scars are the most common exterior platform surface for each raw material. Trimming is most common on non-clast silcrete, followed by clast. 90

Table 8.20. Mean length, width, thickness, platform width and platform thickness of clast silcrete complete flakes in the four assemblages. length F=15.14, *df* 3, 2632, p=<0.001, width F=21.02, *df* 3, 2632, p=<0.001, thickness F=22.42, *df* 3, 2632, p=<0.001, platform width F=11.51, *df* 3, 2376, p=<0.001, platform thickness F=13.57, *df* 3,2376, p=<0.001. 92

Table 8.21. Mean length, width, thickness, platform width and platform thickness of non-clast silcrete complete flakes in the four assemblages. length F=6.07, *df* 3, 536, p=<0.001, width F=5.21, *df* 3, 535, p=<0.001, thickness F=4.28, *df* 3, 536, p=0.005 platform width F=3.43, *df* 3, 469, p=0.017, platform thickness F=2.08, *df* 3, 470, p=0.103). 92

Table 8.22. Mean length, width, thickness, platform width and platform thickness of quartz complete flakes in the four assemblages (complete bipolar flakes not included). length, F=11.78 *df* 3, 7540, p=<0.001, width, F =24.83, *df* 3, 7540, p=<0.001, thickness, F=42.81, *df* 3, 7540, p =<0.001, platform width F=29.37, *df* 3, 6656, p=<0.001, platform thickness F=41.25, *df* 3, 6656, p=<0.001. 92

Table 8.23. Results of least significance difference test (LSD) for complete flakes on different raw materials in the CN1 assemblage. Only those results that returned a significant difference are displayed. 93

Table 8.24. Results of least significance difference test (LSD) for complete flakes on different raw materials in the CN3 assemblage. Only those results that returned a significant difference are displayed. 93

Table 8.25. Results of least significance difference test (LSD) for complete flakes on different raw materials in the KZ1 assemblage. Only those results that returned a significant difference are displayed. 94

Table 8.26. Results of least significance difference test (LSD) for complete flakes on different raw materials in the KZ2 assemblage. Only those results that returned a significant difference are displayed. 94

Table 8.27. Results of least significance difference test (LSD) for clast silcrete complete flakes between assemblages. Only those results that returned a significant difference are displayed. 95

Table 8.28. Results of least significance difference test (LSD) for non-clast silcrete complete flakes between assemblages. Only those results that returned a significant difference are displayed. 95

Table 8.29. Results of least significance difference test (LSD) for quartz complete flakes between assemblages. Only those results that returned a significant difference are displayed. 96

Table 8.30. Clast silcrete platform type frequency and percentage (in brackets) for each assemblage. Uni platforms are the most common type across all assemblages, followed by bi platforms. Cortical and crushed platforms are uncommon. 99

Table 8.31. Non-clast silcrete platform type frequency and percentage (in brackets) for each assemblage. Uni platforms are the most common type across all assemblages, followed by bi platforms, although the proportion of these is greater than that for clast silcrete. Cortical and crushed platforms are uncommon. 99

Table 8.32. Quartz platform type frequency and percentage (in brackets) for each assemblage. Uni platforms are the most common type for quartz, however the proportions of cortical platforms are greater than that for the other raw material types. This supports earlier conclusions that cortical quartz nodules were predominantly reduced at the locations. 99

Table 8.33. Flake to MNT (minimum number of tools) ratio per raw material and assemblage. Proportionally more non-clast silcrete flakes are retouched into tools, followed by clast silcrete in three assemblages (except KZ1 where the quartz ratio is lower). Quartz flakes are generally the least frequently resharpened. 100

Table 8.34. Complete tool form frequency and percentage (in brackets) per raw material type in the CN1 assemblage. Scrapers are the most common tool form within each raw material class. There are less notched artefacts (denticulates and notches) on non-clast silcrete and more utilised artefacts compared to clast silcrete and quartz. 100

Table 8.35. Complete tool form frequency and percentage (in brackets) per raw material type in the CN3 assemblage. Scrapers are the most common tool form within each raw material class. There are less notched artefacts (denticulates and notches) on non-clast silcrete and more utilised artefacts compared to clast silcrete and quartz. 100

Table 8.36. Complete tool form frequency and percentage (in brackets) per raw material type in the KZ1 assemblage. Scrapers are the most common tool form within each raw material class. There are less notched artefacts (denticulates and notches) on non-clast silcrete and more utilised artefacts compared to clast silcrete and quartz. 102

Table 8.37. Complete tool form frequency and percentage (in brackets) per raw material type in the KZ2 assemblage. Scrapers are the most common tool form within each raw material class. There are less notched artefacts (denticulates and notches) on non-clast silcrete and more utilised artefacts compared to clast silcrete and quartz. 102

Table 8.38. Ranking of the proportion of scrapers per raw material and assemblage. A value of one indicates the highest proportion and a value of four the lowest. 102

Table 8.39. Complete flake to complete tool scraper surface area ratio. Quartz has the same value in each of the four assemblages and in some assemblages the value for quartz is greater than that of non-clast silcrete indicating more intensive resharpening. In general, clast scrapers are the most intensively resharpened and both types of silcrete are relatively intensively resharpened in the KZ1 assemblage. 103

Table 8.40. Summary statistics for complete flake and complete scraper surface area. 103

Table 8.41. Mean number of retouched quadrants on complete scrapers per raw material type in the four assemblages. 104

Table 9.1. Artefact frequency and percentage (in brackets) per raw material type in the SQ1 assemblage. 108

Table 9.2. Complete tool frequency and percentage (in brackets) per raw material type in the SQ1 assemblage. 108

Table 9.3. Number of seed grinding artefacts recorded per assemblage for the Pine Point-Langwell study area. Milling slab fragments are the most numerous forms. Also of note is the large number of mortars in the KZ2 assemblage. 109

Table 9.4. Radiocarbon determinations on charcoal from heat retainer hearths at the ND location. (Data provided by Dr. Simon Holdaway and Dr. Patricia Fanning). 113

Table 9.5. Radiocarbon determinations on charcoal from heat retainer hearths at the SC location. (Data provided by Dr. Simon Holdaway and Dr. Patricia Fanning). 113

Table 9.6. Raw material proportion and percentage (in brackets) calculated by number for the Pine Point-Langwell, Fowlers Gap (ND and SC) and Burkes Cave (BC) assemblages. Burkes Cave is somewhat different to the other six assemblages being dominated by silcrete as opposed to quartz. CN1 and CN3 have the largest proportion of quartz and the smallest proportion of silcrete. 115

Table 9.7. Minimum number of flakes per core for each major raw material class in the Pine Point-Langwell, Fowlers Gap and Burkes Cave assemblages. Non-clast silcrete cores are most intensively worked at BC, KZ2, CN1 and CN3, but clast silcrete cores are more intensively worked than non-clast silcrete cores at KZ1, ND and SC. Quartz cores are the least intensively worked. 115

Table 9.8. Non-cortical complete flake to cortical complete flake ratio per raw material and assemblage. Quartz has the lowest ratio for each of the assemblages and only the KZ1 and KZ2 assemblages have proportionally more non-cortical flakes than cortical flakes. Clast silcrete has the highest ratio for the Pine Point-Langwell assemblages, but this does not follow the MNF to core ratio for the CN1, CN3 and KZ2 assemblages. 115

Table 9.9. Non-cortical core to cortical core ratio per raw material and assemblage. Quartz has the lowest ratio in each of the assemblages (except non-clast silcrete at CN3 where the ratio could not be calculated) indicating that there are more cortical than non-cortical cores. Clast silcrete has the highest ratio in the CN1, CN3, KZ1, KZ2 and SC assemblages. The opposite is true for BC and ND, where the ratio for non-clast silcrete is higher than that for clast silcrete. 115

Table 9.10. Flake to tool ratio per assemblage and raw material type. Non-clast silcrete shows the lowest ratio and therefore the most intensive tool production in each of the assemblages except SC. Quartz flakes are less frequently retouched, except for KZ1, where proportionally more quartz flakes are retouched than clast silcrete. 115

Table 9.11. Average number of retouched quadrants on complete scrapers per raw material type and assemblage. Quartz scrapers are generally the most lightly resharpened except in the CN3 assemblage, where the average number of retouched quadrants is lower for clast silcrete. 116

Table 9.12. Complete tool form frequency and percentage (in brackets) for clast silcrete per assemblage. 117

Table 9.13. Complete tool form frequency and percentage (in brackets) for non-clast silcrete per assemblage. 117

Table 9.14. Complete tool form frequency and percentage (in brackets) for quartz per assemblage. 117

Abstract

This monograph presents a theoretical and methodological approach to the investigation of deflated surface stone artefact scatters beyond those that emphasise synchronic behavioural interpretations. The study is undertaken on Pine Point and Langwell Stations, two adjoining pastoral leases south of Broken Hill in arid western New South Wales, Australia. The main objective of the study is to investigate long-term accumulated patterns in stone artefact assemblage composition within archaeological deposits with known occupational chronologies. These are derived from the dating of charcoal from heat retainer hearths. It is argued that the Pine Point-Langwell assemblages represent multiple episodes of accumulation over the last 2,000 years. Therefore, the formation of the Pine Point-Langwell assemblages means they are ideal for the investigation of long-term accumulated patterns.

To analyse the composition of the Pine Point-Langwell assemblages, the concepts of artefact use life, curation, the intensity of raw material utilisation and occupation intensity are used. These permit the investigation of assemblage accumulation as a temporal process. Assemblages are not thought of as synchronic functional sets but rather as the consequence of repeated and discontinuous discard episodes over-time. As occupation intensity increases, so does the intensity of raw material utilisation. Cores and tools will be worked more intensively and assemblages will be dominated by local raw material, as access to distant sources becomes restricted.

Analysis of the composition of the Pine Point-Langwell assemblages indicates both consistencies and inconsistencies in the reduction and utilisation of lithic raw materials. Some of the consistencies are argued to reflect the character and distribution of the wider lithic landscape. In general, there is a distance decay relationship in the reduction of silcrete. This relationship is not evident in all measures of reduction intensity. Variation in measures of core reduction is interpreted to reflect the variable nature of occupation through time at each of the locations in both duration and frequency. Over the time span represented in the Pine Point-Langwell occupational chronology, multiple behavioural patterns result in internal assemblage variability.

Environmental variability may also contribute to the formation of variable assemblage patterns. There is evidence from south western New South Wales for environmental oscillation over the period represented by the occupational chronologies in the Pine Point-Langwell study area. This is interpreted as a possible impulse for the punctuated record of human occupation in the area during the last 2000 years. Hiatuses in the occupational chronology provide further evidence of the variability associated with the formation of the assemblages. Finally, notions of continuity and discontinuity in assemblage formation are explored across the wider region of western New South Wales. Late Holocene assemblages from Fowlers Gap and Burkes Cave are compared to the Pine Point-Langwell assemblages.

It is concluded that the approaches to reconstructing past settlement systems in the Australian arid zone are based on a fundamental misunderstanding of the formation of deflated archaeological deposits. This in turn has led to the development of inappropriate interpretive frameworks. These frameworks often ignore chronology and assume both contemporaniety and consistency in behaviour through time. This denies the opportunity to investigate the diachronic aspects of deflated deposits, both in terms of occupational chronologies and discontinuities in raw material management and reduction.

Acknowledgements

This monograph is a slightly modified version of my PhD. A great many people have assisted me in this process. I acknowledge the assistance of my PhD supervisors Dr. Simon Holdaway and Dr. Peter Sheppard of the Department of Anthropology, The University of Auckland. Dr. Patricia Fanning of the Graduate School of the Environment, Macquarie University has provided valuable assistance over many years. I also wish to acknowledge Simon Holdaway and Trish Fanning for allowing me to access the WNSWAP database including the radiocarbon determinations from the ND and SC locations.

The support of the Broken Hill Local Aboriginal Land Council is greatly appreciated. A special thanks also to Ron and Marilyn Harvy and family of Pine Point Station and Doug and Joy Harrison and family of Langwell Station for allowing the project to be undertaken on their pastoral leases and also to Robert Pearce of Sunnydale Station. Badger Bates formerly of the New South Wales National Parks and Wildlife Service Broken Hill District Office suggested several possible locations for the project to be undertaken and I thank him for his help.

Many of the staff in the Department of Anthropology at Auckland provided assistance including Hamish Macdonald and Tim Mackrell who helped with photography equipment and photo production, and Joan Lawrence and Seline McNamee for drawing the figures. Thank you also to Hamish Macdonald who formatted my thesis into this BAR.

The assistance of the following people (in no particular order) is also acknowledged. Harry Allen, Melinda Allen, Peter White, Robin Torrence, Richard Robins, Leanne Brass and staff of the Australian Museum, Sydney, David Frankel, LuAnn Wandsnider, Jeffery Parsons, the Edwards Family formerly of Broughton Vale Station, John Pickard, Thegn Ladefoged, Rod Wallace, Sarah Martin, Harvey Johnston, Peter Hiscock, Nikki Stern, staff at the Department of Mineral Resources, Broken Hill, staff at the Department of Land and Water Conservation, Broken Hill. Thanks to Bridget Mosley for the very long hours spent in the field during the 2001 field season. A University of Auckland Doctoral Scholarship funded the majority of this research. The University of Waikato Graduate Radiocarbon Dating Fund, the Australian Institute of Aboriginal and Torres Strait Islander Studies and Dr. Simon Holdaway and Dr. Patricia Fanning provided additional funding for radiocarbon determinations. Finally I thank my wife Lisa for her patience and support.

Chapter One: Introduction to the Study

For many archaeologists, the surface archaeological deposits that occur across much of arid Australia represent a tantalising subject for investigation. The character of these deposits as extensive distributions of stone artefacts on lagged and deflated surfaces in a range of environmental contexts appears to support reconstructions of past adaptive behaviours. But these reconstructions lack a consideration of the complex geomorphic and behavioural histories represented in deflated accumulations. Interpretations have been proposed and confirmed without detailed chronological frameworks. The end result is a synchronic behavioural reconstruction, in which change has no significant role. Stasis is emphasised, and when change occurs it is characterised by global (pan-continental) environmental events that result in a large-scale logistical re-organisation of the settlement system (M.A. Smith 1993:47-48). However, as emphasised in some recent studies, e.g., (Fanning and Holdaway 2001a, Holdaway *et al.* 2000, Holdaway *et al.* 2002a, Shiner *et al.* 2005, 2007) this is not the only way to construct a history from the surface deposits of arid Australia. The alternative is not to develop a synchronic perception of the past, but instead to construct one that is capable of investigating issues of change in material drawn from surface deposits at far finer temporal and spatial scales than conventional approaches. What follows is not a transitional journey from predictive model to archaeological interpretation. In the following chapters, a new approach to the writing of prehistory in the Australian arid zone is proposed, an approach that will be unfamiliar to those already acquainted with surface archaeological research in the Australian arid zone. It is an approach that will not make significant contributions to existing models, but rather questions the very foundations upon which existing interpretative frameworks are built.

This research represents a departure from the more conventional settlement-subsistence pattern studies that are the main outcome of Australian arid zone surface archaeology e.g., (Barton 2001, Ross 1981, 1984, Ross *et al.* 1992, M.A. Smith 1988, 1989a, 1993, Veth 1989, 1993). These models are largely based upon ethnographic observations of mobility and resource exploitation (e.g., Veth 1993). Typically, these link patterns in stone artefact assemblage composition to the structure and distribution of key resources, particularly the availability and permanency of water, forming a series of predictive statements regarding the composition and distribution of the archaeological record e.g., (Ross 1981, 1984, M.A. Smith 1989a, 1996, Veth 1989, 1993). To investigate assemblage composition, stone artefacts are studied from a range of different landforms and land surfaces within different ecological settings. Assemblages are compared to reveal patterns in mobility and resource exploitation (often the economics of stone procurement within a distance decay framework of lithic reduction is used as a proxy for mobility). Variation between assemblages is argued to represent functional differences between multiple locations.

These comparisons are based upon two false assumptions. Firstly, it is assumed that archaeological deposits from different landforms and land surfaces have similar chronological contexts and therefore can be regarded as contemporaneous material residues of the same adaptive system (Zvelebil *et al.* 1992). Secondly, that the present environment provides an analogue for the distribution and availability of resources during the past. Behavioural interpretations based on these two assumptions do not take account of the complex formation history of the archaeological record and the variable nature of the environment. Practitioners, who have acknowledged the surface record as a palimpsest of accumulation beyond interpretation at short-term (ethnographic) scales of behaviour e.g., (Barton 2001), have rarely attempted to establish the temporal context of accumulation at individual places. Thus, lithic assemblages are interpreted within very coarse temporal frameworks (often in the tens of thousands of years) in which contemporaneity is assumed rather than demonstrated.

The weaknesses of these approaches highlight the importance of understanding what the formation of surface archaeological deposits represents in terms of temporal resolution. This must be a pre-requisite for behavioural interpretations of archaeological phenomena (Frankel 1993). The temporal resolution of archaeological deposits is dependent upon the rate of cultural and geomorphic deposition (Binford 1982). The highest resolution occurs in deposits that accumulate rapidly. Although in many cases the rate of both sediment accumulation and cultural deposition is intermittent and localised (Stern 1994). The temporal precision necessary for the identification of discrete events in deflated deposits is seldom available (Wandsnider and Ebert 1988). Therefore, most archaeological deposits are time-averaged accumulations of material representing multiple behavioural events (Stern 1993, 1994). Ethnographic and ecological uniformitarian interpretive frameworks of past land use patterns can only be imposed upon this record if the temporal dimension of time-averaged deposits is ignored (Stern 1994).

Unlike settlement-subsistence pattern studies where macro scale comparisons are drawn between assemblages from a range of geomorphic contexts, the present study establishes occupational histories for individual places, thus providing a temporal context for investigating the character of assemblage accumulation. The general aim of this approach is to demonstrate the potential of surface archaeological deposits for investigating the long-term histories of place use (see Wandsnider 1998a) across an area of arid western New South Wales (NSW), Australia. This aim is addressed through the development of an occupational chronology within a geomorphic context for individual places. In the present study this is achieved through the radiocarbon dating of heat retainer hearths.

Recent research in the arid zone of western New South Wales e.g., (Fanning 2002, Fanning and Holdaway 2001a, Holdaway *et al.* 2002a) has established that detailed investigation of the sedimentary record is necessary to understand and interpret the temporal context of the surface archaeological record. Landscapes are in a perpetual state of flux, and the surface archaeological record visible in the present does not mirror that of any stage during the past, but rather reflects the operation of geomorphic processes that have led to the exposure and burial of artefact scatters (Fanning 2002). These processes operate at different spatial and temporal scales, and therefore the contemporaneity of archaeological deposits, especially those from different landforms (geomorphic features e.g. terraces, dunes, floodplains etc.), must be demonstrated rather than assumed.

Establishing a chronological context for surface deposits within a geomorphologic framework is crucial to analyses that compare archaeological materials between multiple locations (Stern 1993). This is particularly true of those that seek to investigate the relationship between assemblage composition, occupation frequency, duration and mobility. For example, a surface with a dense distribution of artefacts may appear to represent a greater intensity of occupation than a surface with few artefacts. However, the sedimentary record and the archaeological chronology may indicate that the first surface is considerably older, potentially representing a longer occupational chronology than the second surface. Older surfaces may also represent the accumulation of artefacts over a greater period of time. For this reason, assemblages that have accumulated over varied amounts of time are not directly comparable within the existing a-temporal frameworks of settlement-subsistence pattern models.

Equally problematic are attempts to define functional links between assemblage composition and the ecological characteristics of site catchment areas (as defined by Vita-Finzi and Higgs 1970). For large parts of the Australian arid zone, the present landscape reflects the past 150 years of pastoral landuse, and is characterised by large-scale changes in hydrology and ecology (Fanning 1994, 1999). The present environment does not reflect that of the past, and is therefore an inappropriate analogue for predicting the distribution and composition of the surface archaeological record across the landscape. Under changing environmental conditions, the nature of land use may alter, resulting in differences in the way that people used places and the rates at which they discarded artefacts. Consequently, static and functionally specific adaptive models cannot be applied to deflated archaeological deposits that represent accumulation over long periods of time. Even in situations where detailed environmental records exist, these are of limited utility in the interpretation of temporal and spatial variation in assemblage composition, if the temporal context of occupation at individual locations is unknown or assumed to be contemporaneous across broad regions.

The alternative approach to current settlement-subsistence models is to investigate the long-term histories of place use within a clear chronological framework that establishes the temporal resolution of archaeological accumulations. To be successful, this approach must go beyond the simple assertion that aggregated deposits are not amenable to interpretations based on short-term behaviour. Additionally, it must offer more than simple synchronic functional-ecological explanations of assemblage composition. The detailed contextualisation of individual locations within spatial, temporal, geomorphic and ecological frameworks is required. Once these contexts are established, the amount of behaviour represented in stone artefact assemblage composition can be compared between multiple locations. These ideas are expanded below.

Research Aims

This research focuses on the 'archaeology of place': more specifically the long-term place use histories of individual locations (Binford 1982, Wandsnider 1998a, 1998b). The assemblages that archaeologists find at places on the archaeological landscape reflect complex occupational histories, ranging from single to multiple occupation episodes of varied intensity and duration (Wandsnider 1998b:98). The study of multiple place use histories establishes the basis for investigations of long-term variability in evolutionary processes (Wandsnider 1998b). Through this approach notions of change and process can be addressed in surface archaeological deposits. Constructing occupational chronologies for different places highlights the variable history of landscape occupation. To address the research aim of this study, the spatial and temporal distribution of surface archaeological deposits at four locations near the confluence of Rantyga and Pine Creeks in the Pine Point-Langwell study area are investigated in detail.

The specific aims of the research are:

- To quantify the composition of lithic assemblages from the Pine Point-Langwell study area that potentially represent the accumulation of artefacts from multiple behavioural events
- To investigate the relationship between assemblage composition and behaviour, especially in terms of raw material transport, raw material reduction, mobility

strategies and occupation intensity in assemblages where both time and access to raw material is controlled

- To incorporate all of the above into a consideration of spatial and temporal variability, with the aim of developing a diachronic history from archaeological deposits traditionally viewed as synchronic and uninformative with regards to notions of change
- To undertake a wider scale investigation of the occupational histories of multiple locations across far western New South Wales, by comparing the compositional of surface lithic assemblages from the Pine Point-Langwell area to those from Burkes Cave and Fowlers Gap

The Study Area

The study area is situated on Pine Point and Langwell Stations, approximately 50 kilometres south of the city of Broken Hill, in far western New South Wales, Australia (Figure 1.1). Pine Point and Langwell are adjoining pastoral leases covering a combined area of 73,000 hectares on the eastern margin of the arid zone. Both stations predominantly run sheep for wool production, with a smaller number of cattle run on Langwell Station.

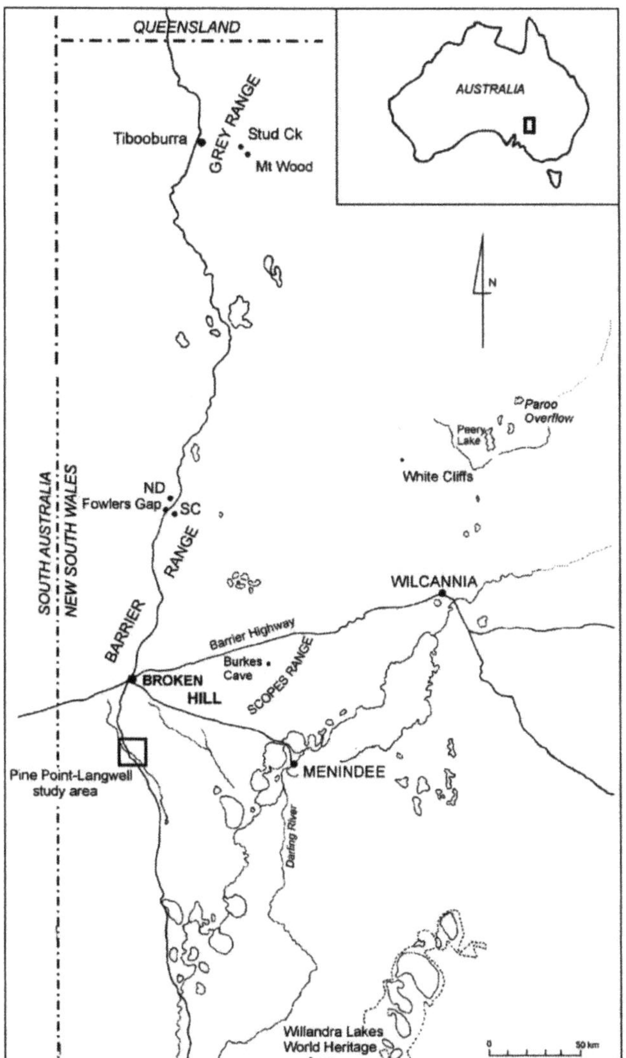

Figure 1.1 Location of the Pine Point-Langwell study area and other features of Western New South Wales mentioned in the text, including the location of assemblages from Fowlers Gap (ND and SC) and Burkes Cave.

The northern section of the study area is situated within the foothills of the Barrier Range, while extending to the southeast is a flat to gently undulating quaternary sand plain. The area is on the edge of the Murray-Darling drainage basin with the Darling River located approximately 70 kilometres to the east. Presently, there are no natural sources of permanent water within the study area. Pine Creek is the major watercourse, with the smaller Rantyga Creek joining Pine Creek on Langwell Station. Pine Creek flows into a small swamp on Kudgee Station, approximately 30 kilometres south of Langwell Station. Smaller ephemeral creeks and drainage lines are distributed widely across both properties, and these often flow into small swamps forming small internal drainage basins. Box swamps (depressions marked by a ring of black box Eucalyptus largiflorens) of varied size are scattered throughout the study area. These rarely contain water, but once full, can retain water for periods of up to 12 months.

Spatially extensive deflated surface archaeological distributions consisting of stone artefact scatters and heat retainer hearths occur on a range of landforms and surfaces. Many of these are concentrated along the margins of Pine and Rantyga Creeks, but these also occur around box swamps and along drainage lines in the hills and the sand plain. Artefact distributions are also found on isolated exposed surfaces away from water retaining features. The condition of deposits varies from those with little gullying to those on steep slopes with significant gullying (erosion) and evidence for dispersion in the form of artefact size sorting and dispersal of heat retainers (Fanning and Holdaway 2001b). Worked outcrops of silcrete also occur on the residual hills in the north of the study area.

Very little archaeological research had been conducted within the immediate context of the study area prior to the present study. During the late 1970's, a portion of the area crossing Pine Creek was surveyed for the Mildura to Broken Hill power line. A number of artefact distributions were recorded on the plains south of Pine Creek and along Pine and Rantyga Creeks (McIntyre 1977, 1981). Other than several contract assessment projects for seismic lines south of the study area e.g., (Martin 1985a, 1985b), archaeological research has been centred around the Menindee Lakes system on the Darling River some 70 kilometres to the east (Aplin *et al.* 1981, Buchan and White 1981, Pretty 1981, C. Webb 1993, 1996). The Menindee Lakes present a very different landscape to that of the present study area and the majority of the research undertaken is not directly relevant to the questions addressed here.

Geomorphic History of Western New South Wales and the Formation of Deflated Surface Archaeological Deposits

The surface archaeological record is formed through the complex interaction of both behavioural and geological processes (Butzer 1994, Schiffer 1972, 1983). Understanding the dynamics of this interaction is central to our ability to develop behavioural interpretations of the archaeological record (Dancey 1998). Acknowledgement of this has resulted in a greater emphasis on investigating the geomorphic characteristics of surface deposits, in particular those post-depositional processes that potentially alter behavioural configurations (Waters 1992, Waters and Kuehn 1996).

The Surface Archaeological Record – What is it?

In arid Australia, the majority of the surface archaeological record consists of distributions of stone artefacts with heat retainer hearths, and occasionally other surface phenomena such as quarries, scarred trees, shell middens, mounds, burials and stone arrangements (Holdaway *et al.* 1998). This study is confined to the investigation of stone artefacts, heat retainer hearths and quarries.

Across arid Australia surface stone artefact scatters occur on a wide range of landforms as spatially extensive distributions (e.g., Davidson 1983, Florek 1987, Hercus and Clarke 1986, Ross 1981, Smith *et al.* 1991). These landforms include quarries on residual surfaces, e.g., hills and rises (Doelman *et al.* 2001), stone artefact distributions spatially associated with mound springs (Florek 1987), alluvial systems (Holdaway *et al.* 1998), and dunes (Smith *et al.* 1991). They are often found on lagged and deflated surfaces where smaller particles, such as sand, silt and clay, have been removed by aeolian and fluvial erosion, resulting in the deflation of heavier clasts, including artefacts. The hearths found on these surfaces consist of heat retainers (either of stone, clay or termite nest) in varying states of dispersal. Heat retainers often cap the bottom of the fire pit where remnant charcoal may be found. Australian archaeologists have adopted several different definitions of quarries (Hiscock and Mitchell 1993:19). The term is used here in a very general sense to describe an outcrop of rock that has evidence of being exploited as a source of stone for the production of stone artefacts. Quarries are widespread in areas where rock suitable for the production of stone artefacts is found (e.g., silcrete, sandstone, quartz and quartzite in Western New South Wales) (Doelman *et al.* 2001). Other types of stone formations e.g., gravel bars in creeks and gibber (stone) pavements also serve as sources.

Geomorphic Processes and the Formation of the Western New South Wales Surface Archaeological Record

The Western New South Wales Archaeological Program (WNSWAP) has investigated the role of geomorphic processes in the formation of surface archaeological deposits across three different study areas; Stud Creek, Fowlers Gap and Peery Lake (Fanning 1999, Holdaway *et al.* 1998, Holdaway *et al.* 2000, Holdaway *et al.* 2002a, 2002b). The results of this research have contributed to a general model of deflated archaeological deposit formation. These concepts can be applied to a broad consideration of surface deposits in other areas of western New South Wales, more specifically the Pine Point-Langwell area.

Recent erosion coupled with a lack of vegetation means that spatially extensive distributions of stone artefacts and heat retainer hearths are visible over much of the arid zone. In western New South Wales, a combination of livestock grazing and vegetation change during the last 150 years has led to widespread erosion in upland areas (Fanning 1999). The effects of these processes have been studied in the Stud Creek catchment in Sturt National Park, near Tibooburra in northwest New South Wales. At Stud Creek, topsoil eroded from the slopes is deposited on floodplains and valley floors, forming a distinctive sedimentary unit referred to as Post European Material (PEM) (Fanning 1999). Stream incision has destabilised the valley floors, resulting in the partial removal of the PEM and the formation of lagged and scalded areas (compact clay surfaces, usually the exposed B horizon of the in situ soil profile, with low infiltration rates) (Soil Conservation Service of New South Wales 1991). During the formation of scalds, artefacts formerly encased within the topsoil matrix deflate onto the scalded surface (Holdaway *et al.* 2000). Other distributions of stone artefacts and hearths remain buried by residual deposits of PEM.

Investigation of the sedimentary history of the Stud Creek Valley reveals that the formation of surface archaeological deposits is characterised by discontinuous geomorphic processes. Five sedimentary units are present in the valley fill of Stud Creek. The uppermost, classified as PEM, was deposited after European occupation of the region. In some locations, the PEM directly overlies brown weakly pedal silty clays, interpreted as representing a former floodplain surface that existed prior to the arrival of Europeans. A sample of sediment from near the top of this unit was dated at 2040 ± 100 years BP (OxL-1050) using the Optically Stimulated Luminescence (OSL) method (Fanning and Holdaway 2001a). In other parts of the valley, this floodplain layer has eroded and the PEM overlies poorly sorted grey/brown gravel and sandy mud deposited in a series of relatively stable pools. Five radiocarbon determinations date this unit to between 4,200 and 4,600 years BP (Fanning 1999; Fanning and Holdaway 2001a). Below are red gravel sands forming an almost continuous unit across the valley. Two accelerator mass spectrometry (AMS) determinations returned age estimates of 5,939 ± 60 years BP (Wk 6237) near the top of the unit, and 12,452 ± 68 years BP (Wk 6236) near the base (Fanning 1999). The gravels suggest a period of greater erosion and deposition in the catchment. The sediments into which the stone artefacts at Stud Creek were deposited, and from which they have eroded, are no older than 5-6,000 years, and are probably as young as 2,000 years. This provides a terminal age for the Stud Creek archaeological deposits (Fanning and Holdaway 2001a).

Below these deposits, red sandy gravels (RSG) have OSL ages that extend from the Late Glacial Maximum to 48.12 ± 2.05 k years BP (OxL-1055) (Fanning and Holdaway 2001a). The most likely explanation for the time depth represented by the RSG unit is that broadly similar sediments were laid down in depositional environments over many thousands of years. A single OSL determination of 75.28 ± 3.50 k years BP (OxL-1056) from below the RSG unit is interpreted as a minimum age for the basal cemented gravely sands.

The sedimentary sequence from Stud Creek is important because it demonstrates the dynamic nature of the sedimentary environment. It is a discontinuous record, reflecting limited sediment supply and dominated by erosion (Fanning and Holdaway 2001a). Unstable conditions predominated over the valley floor from the late Pleistocene through to the mid to late Holocene. It is unlikely that evidence of occupation in the valley prior to the last 2000 years has survived. The discontinuous nature of geomorphic processes means that in areas away from Stud Creek, surfaces and their associated archaeological deposits may show a quite different chronology. For instance, isolated deposits dating to the late Pleistocene have occasionally been discovered in the arid zone e.g., a heat retainer hearth at Lake Yantara 60 kilometres south of Stud Creek returned a radiocarbon determination on charcoal of 26,200 ± 1100 BP (Gak-2121) (Dury and Langford-Smith 1970). Given this temporal variability, it is not possible to interpret different surface scatters as contemporary.

Recent research at Fowlers Gap, 110 kilometres north of Broken Hill (Figure 1.1) provides further evidence of the role of erosion in the formation of surface deposits. Erosion has exposed heat retainer hearths and stone artefact distributions along the margins of upland creeks (Fanning 1999, 2002). Radiocarbon determinations from hearths demonstrate that deposits situated several hundred metres from one another may vary considerably in age. For example, at the Fowlers Creek location at Fowlers Gap, hearths separated by several thousand years are located within several dozen metres of each other.

Implications for the Study of Deflated Surface Distributions

The influence of geomorphic processes on the archaeological record indicates that the concept of assemblages as single depositional events is not applicable to the vast majority of the Australian archaeological record. Both geomorphic and behavioural processes, whether through artefact reuse during subsequent occupations, or from the deflation of artefacts from multiple behavioural events onto a single surface during erosion episodes, result in the reconfiguration of archaeological deposits beyond the scale of single events. While geomorphic processes can distort the archaeological record, or in the case of Stud Creek destroy all archaeological deposits prior to the last 2,000 years, generalisations that all surface deposits are tainted or poor quality sources of information on human behaviour e.g., Mulvaney (1961) are not accurate. Behavioural processes do not distort the archaeological record, but rather it is these along with non-cultural processes e.g., deflation that has led to the palimpsest structure of archaeological deposits (Rossignol 1992, Schiffer 1976), which allow the investigation of accumulated pattern.

There is a long history in Australian archaeology of investigating the relationship between geomorphic processes and the integrity of archaeological deposits e.g., (D. Cameron *et al.* 1990, Hughes 1977, Hughes and Lampert 1977, Richardson 1992, Shawcross 1998, Stockton 1973). The majority of studies have focused on stratified deposits e.g., (Mardaga-Campbell 1986, Theunissen *et al.* 1998) or the relationship between geomorphic processes and artefact breakage e.g., (Hall and Love 1985, Hiscock 1985). Increasingly, the role of geomorphic processes in the formation of surface deposits is the focus of investigation e.g., (Fanning 2002, Littleton 2000, Robins 1999).

Experimental studies are often undertaken to investigate the relationship between geomorphic processes and the structure of the surface archaeological record (e.g., Robins 1999). A limitation of these approaches is the discrepancy between the long periods of time over which multiple geomorphic processes have discontinuously operated across time and space, and the short period of time over which archaeologists can monitor these processes. Experimental studies that attempt to quantify the effects of geomorphic processes over short periods of time are of little value to understanding the influence of these on the formation of the archaeological record over the long-term. Short-term studies reveal little about the rate of spatial disturbance to artefacts because of the episodic nature at which geomorphic processes operate (Fanning and Holdaway 2001b). This is especially true for the arid zone, where long periods of drought are punctuated by large rainfall events.

Settlement Pattern Models and the Surface Archaeological Record of the Arid Zone

Recent studies of the surface archaeological record of the Australian arid zone have largely proceeded without a consideration of the temporal context of the different surfaces upon which hearths and artefacts are distributed. Seasonal settlement-subsistence models are typically the end result of these studies. In the following section, the theoretical background to settlement-subsistence models is reviewed before illustrating the application of these to the Australian arid zone with several examples.

Settlement Pattern Models

Settlement models emerged during the 1950s with the work of Gordon Willey in the Virú Valley of coastal Peru (see Willey 1953). With the ethno-archaeological work of Binford (1978, 1980) and others (e.g., Gould 1991, Yellen 1977) these models have expanded to incorporate aspects of subsistence. Although Australian approaches to settlement-subsistence pattern studies are heavily influenced by Binford's ethno-archaeological work, these have not fully grasped the implications of this research for investigating past settlement patterns.

The relationship between time and assemblage resolution is a major area of departure between Binford's studies and Australian approaches. As assemblages accumulate, resolution at the scale of the event decreases. Therefore, over the long-term, assemblages do not represent the accumulation of artefacts within a yearly adaptive system. A major implication of this is that patterning in the distribution of archaeological 'sites' across arid landscapes does not represent a yearly subsistence cycle of the type described in Binford's Nunamiut ethno-archaeological research. As Binford (1992:52) noted with regard to modern settlement-pattern studies, "The wedding of the assumptions of stylistic seriation and settlement pattern studies into a simple convention for interpreting "site" survey data is sure to result in the invention of a false past" (quotation marks in original). Australian arid zone studies have tended to invent "false pasts" articulated as yearly settlement systems. For example, yearly models of seasonal mobility and resource exploitation have often been imposed upon the archaeological record (e.g., Veth 1993). Even settlement models that envisage changes in land use strategies discuss these within the context of seasonal adaptive strategies (e.g., Thorley 1998a, 1998b, 2001).

Binford (1980) notes two ethnographic types of hunter-gather settlement subsistence systems: forager and collector. These classifications are at the two ends of a spectrum and there is considerable variation in between, but they provide a useful starting point for discussions of mobility and logistical organisation. Foragers are mobile, have a fluid group composition i.e. group size varies according to foraging conditions, do not store food, but gather food on an encounter basis (Binford 1980:7-10). Collectors have a logistically organised food procurement strategy and store food for a least part of the year (Binford 1980:10-12).

Binford (1980) argued that variability in mobility patterns between forager and collector strategies results in variation in the structure and pattern of artefact discard and distribution. Forager strategies generate two types of site: the residential base and the location (Binford 1980:9-10). Variability within forager systems derives from differences in the degree of residential mobility and environmental differences that condition various subsistence activities during the course of the seasonal cycle (Binford 1980). In addition to residential bases and locations, collector strategies generate a range of other special purpose site types including field camps, caches and stations (Binford 1980:10).

Assemblage grain is important to identifying assemblage variation within the subsistence strategies. Fine-grained assemblages are those that have accumulated over short periods of time, in the order of several days (Binford 1980). As the duration of assemblage accumulation increases, the coarseness of the grain increases i.e. the resolution between the assemblage and specific events decreases (Binford 1980:17-19). Increased coarseness reduces assemblage variation between sites of a single or related subsistence system (Binford 1980:17-19). Therefore, through time, assemblages from different parts of the system may become similar in composition.

Binford's categories are derived from short-term ethnographic systems observable at the scale of the yearly cycle. At this scale, assemblage grain is fine compared to archaeological assemblages that have accumulated over the long periods. However, Binford (1980) does not advocate the synchronic application of his observations to the archaeological record. He notes that environmental variation over the long-term would result in changes to redundancy and mobility patterns and that these must be considered before settlement-subsistence systems can be related to the temporal context of archaeological distributions. It is at this point where Australian settlement-subsistence models have departed from Binford's approach.

The Synchronic Functional Paradigm and Site Types

Australian archaeologists have largely viewed site types as functional categories. Although Binford adopted functional terms to describe the location of archaeological sites in his ethno-archaeological studies e.g., residential camp, hunting stand, he acknowledged that the accumulation of archaeological materials at these locations did not always directly relate to the function of the site. As Binford (1978:330) notes:

> Frequently an archaeologist may observe that there is some consistent association between features of the physical landscape and the presence of archaeological sites in a given area. Upon recognition of such a pattern it is not uncommon to study comparatively the artefactual content of the sites as a clue to their "function" and hence a basis for understanding their geographical patterning. The assumption is made that what is in a site betrays the activities conducted there, and that the anticipation of those activities was crucial in selecting the location of the site (italics in original).

Binford's (1978) ethnoarchaeological study of the Mask hunting stand in Alaska demonstrates that there is no simple relationship between function and artefact discard.

He notes that while the intended function of the Mask site was as a hunting stand, the types of activities and more importantly the artefacts discarded there bore little resemblance to this activity. In relation to the Mask site, Binford (1978:360) states:

> The Mask site represents a site where activities conducted are "embedded" in another more basic schedule. None of the activities can be considered "primary" to the mission of the occupants. This means that the specifics of the activities will be largely conditioned by factors other than those which prompted the occupation of the site in the first place.

The artefacts discarded at the Mask hunting stand indicate that a range of activities other than watching game occurred. Clearly, site definitions that are based upon the notion of a single function are not always useful for investigating the composition of many assemblages (Shott 1997). Even those definitions based on multi-functional categories cannot account for the variable range of activities represented in assemblages. This fact has largely been missed by Australian settlement-subsistence pattern studies, where typological site definitions are used to delineate functional variation in the patterning of discard and assemblages across space (e.g., Barton 2001, M. A. Smith 1988, Veth 1993). If the long-term nature of accumulation under varied environmental regimes is considered, then functionally specific assemblages are unlikely to accumulate because environmental oscillation can result in significant logistical reorganisation. Therefore, it is possible that the artefacts discarded at different "hunting stands" may show considerable variability over the long-term as the nature of logistical systems change. In some contexts there may be greater continuity in place use that would lead to similar behaviours, but this would require a degree of cultural and environmental stability.

Site Types in Australian Archaeology

Influenced by ethno-archaeological research (e.g., Binford 1977, Hayden 1977, O'Connell 1977) Australian cultural resource management organisations developed a series of typological site definitions. These are based on the perceived function of the site within a wider settlement system (e.g., Cane 1984). Hiscock and Mitchell's (1993:28) definition of reduction sites, as "the location of early stage stone artefact manufacture (i.e. those stages of manufacture that precede use)" is clearly based on the perceived function of the archaeological materials. While such aspects of stone reduction can be seen in the archaeological record, assemblage composition is often more variable than this. For example, Doelman *et al.* (2001) recorded a range of tool forms in quarry assemblages from northwest New South Wales. Clearly this variability lies beyond the Hiscock and Mitchell definition.

Management decisions are based upon the type, content, age, distribution and preservation of archaeological sites recorded during survey (Pearson and Sullivan 1995:150). These criteria and recommendations regarding the management of archaeological deposits are evaluated within a framework of scientific significance assessment that investigates the representiveness of different types of sites within a given area (Bowdler 1984, Flood 1984, Hiscock and Mitchell 1993, Pearson and Sullivan 1995, L. Smith 1996). Rare site types are often assessed as having a higher scientific significance than more common sites. Some archaeologists (e.g., Bowdler 1984, Robins 1984) have pointed out problems in this framework. For instance, the criteria used to assess scientific significance, e.g. site size, preservation, representiveness and information potential can change over short periods of time in accordance with theoretical and methodological changes in archaeology, or even the amount of fieldwork undertaken.

A major problem with typological site definitions is that like all typologies, these fall foul of variation (N. David and Sterner 1999:97). Variable assemblage content means that typological site definitions are not always broadly applicable. As Cane (1984) discovered during his ethnographic study of stone artefact assemblages in Central Australia, correlating assemblage content with site function is rarely straightforward. Cane (1984:233-234) argued that ethnographic observations indicate that assemblages with high proportions of scrapers represented woodworking locations. But Cane's own investigations demonstrated that high proportions of scrapers were also recorded at sites where ethnographically woodworking was not the primary activity (Cane 1984:217,223,233). This casts further doubt upon the legitimacy of inferring functional correlation based upon artefact frequency.

Adams and Adams (1991) note that the definition of a typology based on function is not straightforward, and that some variation can be subsumed in definitions based on morphology. They (1991:285) argue that "...types that were differentiated at least partly on morphological grounds are grouped, for descriptive or interpretative purposes, into larger taxa on the basis of a presumed affinity of function...". This argument is circular, and is complicated by the difficulty of assigning an object or site to a specific type when it may meet the criteria of several.

In a more general criticism, Wandsnider and Ebert (1988) questioned the utility of site type definitions for investigating issues in settlement archaeology. They argue that site types are difficult to define and contribute little to understanding the broader landscape context of archaeological phenomena. Sullivan III (1992) expands this notion in relation to units of observation and analysis by arguing that typological site definitions are interpretative constraints that obscure the historical data that archaeologists require for the investigation of evolutionary problems. Despite these criticisms, and recent acknowledgement of the long-term character of assemblage accumulation, functionally defined site types remain a staple of Australian arid zone archaeological studies.

Arid Zone Settlement-Subsistence Models

Settlement-subsistence models are the dominant interpretive paradigm in Australian arid zone archaeology. Most of these are framed as rigid synchronic predictions usually derived from ethnographic observations of mobility, occupation and resource use. They are largely based on notions of continuity and cultural stability.

In a significant study, Veth (1993, 1996) proposed a settlement-subsistence model for the arid Rudall River region of Western Australia. Using ethnographic information and an investigation of the archaeological record Veth (1993:78) argued for a seasonal pattern of resource exploitation, mainly influenced by rainfall and a flexible cycle of aggregation and dispersal. Following summer rainfall, mobility is high with small land using groups occupying widely scattered ephemeral water sources. Sites in these areas are referred to as satellite camps and occupation is opportunistic and short-term (Veth 1993:83). As ephemeral water sources dry, groups gradually move back into core habitation areas where more permanent waters occur. Occupation duration is greatest at sites adjacent to more permanent water sources, referred to as core habitation camps (Veth 1993:83). Veth argues that stone artefact assemblage composition reflects the relationship between occupation duration and water permanency, and that the most intensively reduced assemblages occur at core habitation sites.

Veth's predictions about mobility and occupation duration being tied to water permanency in an arid landscape are entirely reasonable. However, the model is developed and applied within a synchronic framework based on the notion that human behavioural-environmental interaction remained stable for long periods of time. Veth (1993:80) states "...to legitimately apply the predictive model here, within an archaeological context...it is necessary to assume that the settlement-subsistence system has remained broadly similar since the earliest demonstrable date for occupation; in the study area this is 5,000 BP...". The model is dependant upon the assumption that the environment remained stable and that the present environment provides an analogue for the past. Reviewing the environmental history of the study area, Veth (1993:9) notes considerable environmental variability during the last 7,000 years, including a major dry period between 4,000 and 2,000 BP. Therefore, present environmental conditions e.g., rainfall averages reveal little about the environmental conditions of the Rudall River region during the period over which the archaeological record accumulated. Furthermore, increased sediment yields in arid zone drainage systems during the period of pastoral land use have resulted in significant hydrological changes (Pickard 1994).

Table 1.1 Radiocarbon determinations from the four rockshelters in Veth's Rudall River study (Adapted from Veth 1993:Chapter Five). The wide range of determinations between the rockshelters indicates that basal occupation was not contemporaneous. Both Karlamilyi and Jalpiyari have evidence of use far earlier than Winakurijuna and Yulpul, therefore as a group the rockshelters do not form a single settlement system.

Site	Lab ID	Result	Depth cm
Karlamilyi	WK-1092	1120 ± 50	58-60
	WK-1093	3180 ± 70	117-120
Winakurijuna	WK-1158	900 ± 70	43-45
Yulpul	WK-1256	315 ± 150	35-37
Jalpiyari	WK-1288	1020 ± 50	40-42
	WK-1255	5030 ± 60	75-77

The chronological context for Veth's model is derived from the excavation of four rockshelters. The number of radiocarbon determinations per rockshelter varies between one and two and these are mostly obtained from near the bottom of the deposit (Table 1.1). Comparison of the basal dates of the four rockshelters (Veth 1993:45,55,60,63) reveals that occupation of each was not necessarily contemporaneous. The basal determinations at two of the rockshelters (Winakurijana and Yulpui) are considerably more recent than those from Karlamilyi and Jalpiyari. It can be argued that three of the determinations (WK-1092, WK-1158, and WK-1288) are broadly contemporaneous, but the basal determinations from Karlamilyi, Yulpul and Jalpiyari vary considerably indicating that at least some occupations where not contemporaneous.

Surface scatters of stone artefacts are a major analytical component of the model, but no attempt is made to establish a chronological context for these (see also Gould 1995:563 for a criticism of this). Within the model, stone artefact assemblages from different surface deposits are directly compared to one another and to the rockshelter assemblages without a consideration of the temporal context of artefact accumulation. There is no justification for assuming that archaeological materials from different locations belong to the same historically related adaptive system (Bird and Frankel 1991a). Instead, assemblages may represent accumulation from multiple systems, both at inter and intra site scales.

Further problems with Australian arid zone settlement pattern models are illustrated in more recent research. Thorley (1998b) documents many of the problems of synchronic models in his doctoral research in Central Australia. He attempted to address some of these through the development of a six-phase diachronic settlement model spanning 30,000 years BP to present in the Palmer River Catchment. Within this timeframe, he (1998b:311) argued that data resolution increases through time, but that it remains too broad for the identification of discrete occupation events. Importantly, he acknowledges that yearly or longer environmental fluctuation will lead to variation in settlement-subsistence systems. Archaeologically this can be seen as discontinuities in occupational chronologies resulting from abandonment.

The archaeological evidence for the Palmer River catchment is derived from the excavation of five rockshelters, a single stratified open deposit and the survey of numerous surface artefact distributions. A total of 16 radiocarbon determinations were obtained from the five rockshelters and the open stratified deposit (Thorley 1998b:220). Of these, seven determinations are from the Kulpi Mara Rockshelter, with the earliest approximately at 30,000 years BP and the most recent at 2,500 years BP. Three of the other rockshelters have determinations at approximately 3,000 years BP, the other rockshelter at 1,500 years BP and a determination of approximately 500 years BP was obtained for the stratified open deposit (Thorley 1998b:220). In addition, a total of 71 surface artefact scatters were recorded as sites (Thorley 1998b:112).

Thorley (1998b) acknowledges that establishing the chronological context of surface deposits is important to investigating the archaeological record. However, despite the widespread presence of hearths on land surfaces with stone artefact distributions (Thorley 1998b:356), none of these were dated to establish occupational chronologies for the surface deposits. Based on a limited geomorphologic study and artefact typology, Thorley (1998b:287-288) concluded that the majority of surface artefact distributions occur on surfaces dating to less than 500 years BP. The typological evidence consists of the low number of backed blades recorded in surface assemblages compared to rockshelters assemblages, where backed blades occur in greater frequencies in deposits older than 500 BP (Thorley 1998b:288). On this basis the majority of surface deposits, which were recorded in a range of landscape contexts, were treated as broadly contemporaneous (Thorley 1998b:309). Thorley (1998b:11) argued that the study of settlement patterns must be undertaken through the investigation of "...different, yet related, sites". How are 'related' sites defined? Separate archaeological deposits can be considered related in one context i.e. spatial, for deposits within a single catchment area, but may be unrelated in another context, i.e. temporal if the deposits within the catchment are all different ages. These are scale dependent considerations, and they have important implications for the development of behavioural models. Thorley (1998b:11) contends that the surface deposits in the Palmer River catchment are related because they are contained within a single drainage system and have a similar antiquity. Neither of these two arguments is convincing. Firstly, co-location in space does not imply a historical relationship between different archaeological deposits. As demonstrated by hearth-dating in western New South Wales (Holdaway *et al.* 2002a), surfaces separated by several hundred metres or less may represent numerous accumulation episodes over vastly different periods of time. Secondly, studies of occupational chronologies for surface deposits indicate that continuity and discontinuity in occupational histories can be defined at temporal scales finer than 500-year units (e.g., Fanning and Holdaway 2001a, Holdaway *et al.* 2002a). Therefore, if it is accepted that the majority of surface deposits in the Palmer River catchment are less than 500 BP in age, then it is possible that significant spatial and temporal variation in occupational histories will exist within these units. The probability of variation in occupational histories increases if the surface deposits represent accumulation over a longer period of time.

The assumption that surface deposits are "broadly contemporaneous" (e.g., Thorley 1998b:309) simplifies the temporal complexity of the archaeological record. The temporal context of surface deposits cannot be understood through generalisations, especially when these are applied to dozens of locations distributed over a range of geomorphic contexts. Small-scale micro-geomorphologic variation (see Fanning and Holdaway 2001a) plays as important a role in structuring the temporal distribution of archaeological deposits as past human behaviour. Absolute chronological histories must be established within geomorphologic frameworks at individual locations. Only then can temporal contemporaneity be inferred between multiple and spatially dispersed archaeological deposits.

Settlement-subsistence models developed within a synchronic framework reveal little about the nature of human adaptive systems. As Dewar and McBride (1992:228) note "Attempts to compare site distributions to models of an annual subsistence round are quite simply attempts to compare radically different phenomena". In particular, functional interpretations of assemblage composition dependent upon assumptions of environmental stability do not reveal the behaviours that govern artefact discard and assemblage accumulation through time.

Predictive Models

Settlement pattern models are crucial to the development of predictive models in Australian archaeology. These are widely applied to identify areas of the landscape where certain site types are likely to occur. Predictive models are based on prior knowledge of archaeological distributions and their relationship to environmental variables. This knowledge is gained through the archaeological survey of sample landscape units. This information is then used to predict the likely distribution of archaeological materials across areas where little or no survey work has been undertaken. At one level, environmental variables can be used to predict the presence of some archaeological distributions. For example, the distribution of stone sources suitable for the manufacture of stone artefacts will only occur in those areas of the landscape with a certain type of geology. Simplistic environmental correlation ignores both the temporal characteristics of the archaeological record and environmental change. Indeed, as Holdaway *et al.* (2002a) suggest, fine scale environmental fluctuation may have important implications for past patterns of landscape use. Environmental fluctuation (of varying scales) that alters the availability and distribution of key resources may result in the abandonment of one area in favour of another (Tomka and Stevenson 1993:192), thus altering the tempo at which the archaeological record accumulates (Stevenson 1982). Environmental fluctuation means that functionally synchronic adaptive land use systems are unlikely to be represented in arid zone archaeological assemblages. As Nelson (2000) points out, abandonment is a process, not an event. Therefore, discontinuity in occupational chronologies should not only be viewed as the passive effect of abandonment (Nelson 2000).

The factors governing patterns in the distribution of archaeological materials are far more complex than environmental correlation alone and, as argued, these may vary through time. In this regard, predictive models can also be criticised from a taphonomic perspective. Most predictive models are based upon an unsophisticated understanding of the influence of geomorphologic processes on archaeological distributions. Geomorphologic processes operate at a range of spatio-temporal scales; in some situations these may completely remove or obscure the archaeological record from large areas of the landscape (A. Brown 1997, Fanning 2002). Without understanding geomorphologic processes at this level, it is problematic to predict the location of archaeological materials across the landscape (Waters and Kuehn 1996).

The sedimentary record from the Stud Creek valley is a clear example of this. Geomorphic processes have removed any archaeological deposits in the valley older than 2000 years BP (Fanning and Holdaway 2001a). Adjacent drainage systems to Stud Creek may exhibit different sedimentary histories and these will have implications for the ages of the archaeological deposits that occur within them. Archaeological deposits in the Sandy Creek Gorge on Fowlers Gap Station approximately 200 kilometres south of Stud Creek date to within the last 900 years BP (Simon Holdaway personal communication). The sedimentary record indicates that the Sandy Creek Gorge was washed-out by a large flood event at approximately 1100 years BP (Jansen 2001), removing archaeological deposits before this period. Radiocarbon determinations from heat retainer hearths from other locations on Fowlers Gap indicate human occupation prior to 900 years BP (Simon Holdaway personal communication). Therefore, predicting the spatial and temporal distribution of archaeological deposits in the arid zone is a complicated process that requires confirmation through detailed geomorphic studies and archaeological dating programmes.

In summary, both settlement-subsistence and predictive models marginalise variability through a misunderstanding of the temporal context of the archaeological record. This is especially pertinent to attempts at modelling settlement-subsistence pattern systems where ethnographic systems are directly fitted onto the archaeological record. The surface archaeological record is a complex temporal and spatial patchwork, and understanding the formational history of the record at individual places must precede higher-level comparisons of archaeological content.

An Alternative to Current Settlement Pattern Models – The Pine Point-Langwell Approach

In the previous section it was argued that Australian arid zone settlement pattern studies have paid little attention to the establishment of occupational chronologies for surface archaeological distributions. This has raised a number of issues regarding historical relationships between spatially dispersed archaeological deposits. While the chronological resolution is not presently available for the identification of fine scale temporally associated events between multiple deposits, the establishment of coarse occupational histories based on absolute dating techniques provides a temporal justification for investigating and comparing the accumulation of materials at different places across the landscape. These comparisons are not perceived as observations relevant to individual settlement systems (Dewar and McBride 1992), but rather aim to study the occupational histories of individual places within a framework that focuses on accumulated long-term patterns in artefact discard and reduction.

Geomorphic Processes and the Surface Archaeology of the Pine Point-Langwell Study Area

Previous geoarchaeological research in the rangelands of western New South Wales (e.g., Holdaway *et al.* 2000, Holdaway *et al.* 2002a) helps to develop a general model of archaeological deposit formation for the Pine Point-Langwell study area. Although the present study area is in a different landscape context to the rangelands, a similar range of geomorphic processes are operating. Many areas of Pine Point-Langwell exhibit the deflation of sediment arising from erosion, particularly along the margins of creeks and drainage lines. This has exposed stone artefact distributions and heat retainer hearths forming palimpsest deposits over areas of discontinuous bare ground exceeding 10,000 square metres.

Sediment deposition is also occurring, particularly on the floodplain of the two major creeks in the study area. The present floodplain of Pine Creek is covered in loose sandy sediment that has obscured any artefact distributions that may have been visible. In other areas, particularly the low hills in the north east of Pine Point, residual surfaces contain rock outcrops that form potential sources of material for the manufacture of stone artefacts. The contribution of erosion to the formation of the surface archaeological deposits in the Pine Point-Langwell study area indicate that these represent palimpsests similar to those identified elsewhere (e.g., Holdaway *et al.* 2000).

Theoretical Approach – Place Use History

The episodic and intermittent nature of the formation of surface deposits means that these are not amenable to interpretations based upon short-term behaviour (see Fanning and Holdaway 2001a, Stern 1993, 1994, 1995). Instead, we can consider the role of long-term place use in the structuring of the surface archaeological record (Camilli 1983, Wandsnider 1998b). Occupational histories (comprised of but not limited to duration, frequency, activity and intensity of activity) from places representing accumulation over long periods of time permit investigations of landscape use. The use of places during the past and the number of separate occupations comprise the occupational history (Camilli 1983). Understanding the temporal dimensions of the formation of surface archaeological deposits is crucial to investigating place use histories. Central to this is the establishment of occupational chronologies for archaeological deposits from different parts of the landscape. The comparability of assemblages is dependent upon establishing these, especially for considerations of land use histories.

Schlanger (1992) proposed the term 'persistent place' to describe those areas of the landscape that are the focus of repeated activity through time. The notion of persistent place is useful for investigating the long-term place use histories of individual locations because it acknowledges the role of multiple behavioural events in the accumulation of the archaeological record. Within this context, assemblage accumulation represents multiple processes that may have very different temporal trajectories, rather than the result of a synchronic functional-environmental relationship between discard and place use. The composition of assemblages and the spatial structure of artefact distributions reflect the long-term repeated use of locations.

To investigate the temporal character of assemblage accumulation, the life histories of individual artefacts need to be considered in terms of their probability of discard. To a certain degree, stone artefact assemblage composition reflects the nature of occupation duration (Kuhn 1991:77). Generally, in flaked stone assemblages, curated artefacts are those that have long use-lives (heavily resharpened tools and sometimes cores), while non-curated artefacts have short use-lives (lightly and non-resharpened artefacts) (Shott 1989). Greater amounts of time can be represented in the manufacture and reduction of curated artefacts, and these are more likely than unretouched blanks to be transported between multiple locations. Curated artefacts have a lower probability of discard at any time compared to non-curated artefacts because they remain within the cultural system for a greater amount of time (Holdaway *et al.* 2000). For curated artefacts to be discarded at a given location requires that either occupation duration exceeds artefact use-life, or that tools are nearing the end of their use-lives when occupation commences (Shott 1995). Assemblages with large numbers of long use-life artefacts are indicative of intensively occupied persistent places.

Equally important, but often ignored in Australian arid zone surface archaeology, is the spatial structure of artefact distributions (exceptions are Holdaway *et al.* 1998 and O'Connell 1987). In the arid zone, surface artefact scatters appear as spatially extensive distributions of varying densities over tens of thousands of square metres. Holdaway *et al.* (2000) argued that the temporal character of artefact discard is further represented in the spatial patterning of artefact distribution. At Stud Creek for example, Holdaway *et al.* (2000) demonstrated that highly curated artefacts have a clustered distribution and that the least curated artefacts have a random distribution. This is interpreted as a function of time because inhabitants are drawn to particular geographic features more frequently. At Stud Creek the artefact clusters are on the valley floor adjacent to a chain of former waterholes, now in-filled with sediment (Holdaway *et al.* 2000).

Methodological Approach

Assemblages of several hundred or even one thousand artefacts are not adequate for identifying patterns in assemblage composition. Small assemblages are susceptible to a sample size effect, where diversity measures reflect the small size of the assemblage rather than the true absence of specific items (Hiscock 2001). This is especially a problem if assemblages of different sizes are compared e.g., assemblages with several hundred artefacts compared to assemblages with several thousand. Also, the range of technological analyses that can be undertaken is limited in small assemblages because these lack sufficient numbers of each artefact class. Small assemblages may also result in different raw materials being combined to increase the number of artefacts for some analyses (e.g., Barton 2001), thus ignoring variability in reduction strategies between different stone types. Studies of assemblage composition in other areas of the arid zone (e.g., Holdaway *et al.* 2004, Shiner *et al.* 2005) have identified significant variability in the degree of reduction between different raw material types. Combining different raw materials also compromises the analysis of relationships between local and non-local raw material use. To avoid the effects of outliers on analyses of central tendency and to identify trends in composition and reduction strategies, assemblages of several thousand or more artefacts are required. This is the minimum assemblage size analysed in this study.

Excellent conditions of surface exposure in the arid zone mean that artefact distributions are visible over large areas. In many cases, these deposits represent multiple discard events and over-printing of material (Holdaway *et al.* 2002a). Therefore, artefact distributions appear as a veneer of varied densities across large areas. Ethnoarchaeological and archaeological studies of arid zone artefact distributions suggest that the identification of spatial patterning in low density dispersed distributions requires the investigation of areas greater than 10,000 square metres at individual locations (Holdaway *et al.* 2000, O'Connell 1987). A transect several metres wide or a 100 square metre sampling area will not achieve this. In most cases, spatial studies take the form of transect surveys designed to assess artefact distribution in relation to environmental variables (e.g., Robins 1993, Thorley 1998b, 1999). Transect surveys reveal little about the distribution of stone artefacts at individual localities, not to mention the difficulties of establishing a temporal context for multiple archaeological deposits from a range of geomorphic situations. Furthermore, artefact visibility is largely controlled by micro-geomorphic context.

Although the spatial characteristics of artefact distributions are not the primary concern of the present research, as discussed above previous research indicates that these are important factors that must be considered when recording artefact distributions. Differential visibility in particular can influence the overall composition of assemblages. For instance, sand and vegetation may more readily obscure smaller artefacts than larger artefacts, thus affecting calculations of average artefact size. To circumvent these problems artefact distributions were recorded across areas in excess of 5000 square metres at each of the four study locations considered here. In addition, the micro-geomorphic context of the surface upon which artefacts were recorded was described in one square metre areas.

An archaeological chronology for surface archaeological deposits can be established through from radiocarbon determinations obtained from heat retainer hearths. This has seldom been undertaken despite the fact that heat retainer hearths are common to surface deposits recorded during traditional site based settlement pattern surveys. As Fanning and Holdaway (2001a:101) suggest, obtaining a large number of radiocarbon determinations from heat retainer hearth permits the study of processes thorough the investigation of the combined temporal pattern, rather than viewing determinations in isolation. It is at this level that an occupational chronology will be established for the surface deposits in the Pine Point-Langwell study area. This is based upon radiocarbon determinations from 16 heat retainer hearths.

Summary

With the benefit of a better understanding of deposit formation due to the many local and international studies undertaken over the last decade combined with the technological means to record artefact distributions over large areas, it is easy to highlight the flaws, rather than the contributions, of previous studies. These studies have made a valuable contribution to understanding the composition of the arid zone archaeological record. The theoretical review presented in this chapter has highlighted many of the gaps in arid zone studies. Consideration of these studies within the historical development of Australian archaeology indicates that they were in some cases insightful for their time. However, recent theoretical and methodological developments in some areas of Australian archaeology e.g., the work of WNSWAP indicate that a review of the study of arid zone surface archaeology is over due. The following chapters attempt to go some way towards achieving this review.

Deriving behavioural inferences from the arid zone surface archaeological record is difficult. Surface deposits are palimpsests that contain few features, are difficult to date and consist primarily of spatially extensive distributions of stone artefacts without clear boundaries (Holdaway *et al.* 1998). Establishing the occupational history of individual places within geomorphic contexts is crucial to the development of behavioural interpretations. Once a chronological context is established, the long-term place use histories of locations can be investigated from the perspective of artefact discard, not in terms of a functional-ecological relationship, but rather as a proxy for understanding the duration and intensity of occupation. At a wider landscape scale the long-term histories of place use can be compared between locations with similar chronological contexts to investigate the nature of landscape occupation within specific temporal units. Assemblages with different chronological contexts can be compared to investigate temporal and spatial variation in assemblage composition and reduction through time. This in turn can be related to investigations of change, and diachronic history. Missing from this approach is the notion that settlement-subsistence systems are empirical entities that can be reconstructed from an investigation of the archaeological record. In the next chapter the notion of the surface archaeological record as a palimpsest is considered in more detail, as are the implications of this for interpreting the record in behavioural terms, especially those that relate to issues of variation across space and time. Finally, the implications of these for interpreting the spatial structure of surface artefact distributions are considered.

Chapter Two: Time and the Temporality of Artefact Accumulation

Introduction

In the previous chapter two common perceptions of the surface archaeological record were discussed. Firstly, that surface deposits represent synchronic behaviour, in which materials were deposited at places during the course of similar activities commensurate with ethnographic scales of behaviour. Secondly, deflated deposits represent palimpsests of material from multiple events and that discontinuities in occupational chronologies demonstrate irregularities in land use strategies through time. The former view is a clear misunderstanding of the formation of the surface record, which rarely if ever represents the accumulation of materials from a single event. The latter is viewed as a weakness of the surface record by archaeologists who seek to reconstruct single or short duration events commensurate with ethnographic scales of time (Murray 1993). Surface deposits are perceived as a poor record of human behaviour because geomorphic processes and overprinting from repeated occupation compromise the spatial and temporal integrity of archaeological materials (Sullivan III 1998).

The reality is that much of the surface archaeological record documents the accumulation of materials from multiple behavioural episodes occurring over long periods of discontinuous time (Ebert 1992, Edwards 1995, Holdaway *et al.* 2000, Stern 1993, 1994, Stern *et al.* 2002, Wandsnider 1998b). As discussed in the preceding chapter, the discontinuous nature of deposition means that the record cannot be partitioned into functionally specific assemblages and locations (sites) representative of a past settlement system. Therefore, the characteristics of place use and the types of activities performed there can change through time. As such, archaeological deposits represent accumulation beyond the scale of ethnographic time and are not a direct analogue for the behaviours (in terms of logistical systems) that have resulted in their formation. Australian archaeologists have largely ignored the disparity between the temporal depth of the archaeological and ethnographic records (H. Allen 1996).

Uniformitarian principles permit extrapolation from the short-term to the long-term and have been crucial to explanations of the archaeological phenomena found in the Australian arid zone. However, the way that many Australian archaeologists have applied ethnography to the interpretation of the archaeological record has largely arisen from a misunderstanding of uniformitarian principles. Bailey (1983) noted that there are two distinct ways in which archaeologists have applied these principles. Methodological uniformitarianism involves the application of knowledge gained from the observation of processes in the present to the reconstruction of the past within theoretical frameworks. But this is not the simple extrapolation of modern processes onto the archaeological record as an explanatory device for the same phenomena. Within the application of these processes it is acknowledged that variability exists between the archaeological record and the ethnographic (Bailey 1983:175). As discussed, Binford's study of formation processes at the Mask hunting stand identifies some of the behaviours that lead to artefact discard, but these alone cannot account for the factors (cultural and non-cultural) that result in the assemblages studied by archaeologists. Substantive uniformitarianism uses contemporary processes observed in the present to explain those that operated in the past, often through extrapolation of these onto the archaeological record. An error of substantive uniformitarianism is the assumption that there is a level of consistency between processes observable in the present and those operating during the past, i.e. that they are similar, the same or related (Fletcher 1992:44). If this relationship between processes is not constant, then a range of inappropriate uniformities will be imposed upon the past.

It is argued here that the so-called palimpsest problem is perceived rather than real, since it is founded upon a misunderstanding of the potential of the surface record to inform archaeologists about long-term trends in the accumulation of materials and the use of place/s. One of the major goals of archaeological inquiry is to assess change through time in material culture. In stratified deposits this can be achieved through the laws of superposition, which allows the comparison of different layers. For deflated surface deposits, where stratigraphic relationships have been removed, identifying change through time is more difficult because relative chronological relationships have been removed through erosion. A common approach is to assume that surface archaeological deposits share similar chronological histories to stratified deposits (e.g., Thorley 1998b, Veth 1993). In some cases this may be true, but contemporaniety must be demonstrated rather than assumed. In the Australian arid zone, this method is often used to justify models of late Holocene intensification, where an apparent increase in surface 'site' numbers and discard rates is argued to represent more intensive resource exploitation, decreased mobility and population increase (e.g., Lourandos 1985, 1997, Veth 1993, 1996).

Stratigraphy is not crucial to investigating change, but to undertake such an investigation requires a re-consideration of how change is defined, and at what temporal and spatial scales is it identifiable. The development of place use histories is one method to access change because a major component of these histories is the establishment of occupational chronologies for surface deposits. Occupational chronologies provide the framework for assessing temporal variation in assemblage composition, not necessarily within a single archaeological distribution, but more readily between assemblages from different parts of the landscape. Where there is no culturally defined historical relationship between assemblages, the landscape is the common link and diachronic variability across space the focus of investigations of change. The landscape shows both continuity and discontinuity in characteristics that can vary over both short and long time periods.

Notions of time, the temporal character of assemblage accumulation and the implications of the surface archaeological record as a palimpsest for interpretations of archaeological phenomena are discussed in this chapter.

Chronological Contexts and Surface Archaeological Deposits: A Case Study of Stud Creek

The previous chapter introduced the idea that the dating of surface archaeological features is crucial for understanding both the formation of the record and the development of behavioural interpretations of archaeological phenomena. Few concerted attempts to establish chronological contexts for this type of deposit have been made. Where occupational chronologies from surface contexts have been established, these are more often derived from stratified open deposits (e.g., Allen 1972, Thorley 1998b). This has contributed little to the establishment of archaeological chronologies for lagged surface deposits, but represents a positive step towards acknowledging the importance of temporal context in arid zone surface archaeological studies. This section deals with absolute dating methods only. The problems of using stone artefact typology to date surface deposits are explored more fully in Chapter Three.

Dating Arid Zone Surface Archaeological Deposits

Many of the principles of dating Australian surface deposits were developed during archaeological investigations in the early 1970s at Lake Mungo in western New South Wales. A combination of thermoluminescence and radiocarbon techniques established chronological contexts for stratified and non-stratified archaeological deposits within geomorphologic frameworks of landscape evolution (Barbetti and Allen 1972, Bowler *et al.* 1970, Bowler *et al.* 1972). Aspects of these techniques were applied in other areas of Australia (e.g., Smith *et al.* 1991, Veth *et al.* 1990), but chronologies for surface deposits have only rarely been developed within the detailed geomorphologic contexts comparable to Lake Mungo (e.g., papers in R. Jones *et al.* 1985).

The majority of chronological investigations of surface deposits are directed towards identifying the oldest date for occupation of an area rather than establishing a chronological context for human occupation. Smith *et al.* (1991) for example report five radiocarbon determinations ranging from 14,500 BP to 2,500 BP from heat retainer hearths at the JSN location in the Strzelecki Desert. They focus on the several terminal Pleistocene determinations, largely ignoring the late Holocene determination and the information it yields for understanding the broader occupational history of the JSN locality. Instead, they state that the Pleistocene determinations indicate that JSN was occupied on more than one occasion (Smith *et al.* 1991:190). Little attention is given to the concept that the JSN assemblage represents a palimpsest. The only exploration of this issue is the conclusion that typologically, the majority of artefacts appear to belong to the late Holocene and that few artefacts have weathering characteristic of old assemblages (Smith *et al.* 1991:184, 190).

Smith and Clark (1993:121) note that isolated radiocarbon determinations from individual hearths are of limited value because these provide limited information on the temporal character of occupation. Williams (1988, 1998) discusses radiocarbon determinations obtained from heat retainer hearths at several locations in the Cooper Basin of northern South Australia. But, the greatest number of determinations from any location is two and there is little attempt to establish a chronological framework for individual locations. Bourke (1997) reports two determinations from hearths in the Eastern MacDonnell Ranges of Central Australia. Veth *et al.* (1990) report two terminal Pleistocene radiocarbon determinations from heat retainer hearths adjacent to Cooper Creek in northern South Australia. They contend that the dates indicate probable occupation of the area during the terminal Pleistocene. Beyond this, no attempt is made to establish a chronological framework for the many surface deposits recorded during the survey. Nor is the relationship between the geomorphic history of the landscape and archaeological resolution considered. Despite this, Veth *et al.* (1990:63) argue, based on artefact typology alone (the large number of deposits containing artefacts belonging to the Australian small tool tradition), that the majority of archaeological deposits date to the late Holocene. This assumes that artefact discard did not commence at many of the deposits until the late Holocene, despite the evidence for probable Pleistocene occupation. In the absence of detailed geomorphic studies of landscape formation and the dating of further hearths it is impossible to assess the validity of this statement.

At the Balcoracana Creek locality near the Flinders Range in South Australia, Lampert and Hughes (1988) argue that an age range for carbonate-encrusted stone artefacts on a deflated dune blow-out can be established by thermoluminescence dating of the carbonate rich sands from which the artefacts are presumed to have eroded. Non-carbonate encrusted artefacts within the same deposit are thought to have eroded from more recent overlying sands (Lampert and Hughes 1988:148). Furthermore, they reason that the micro geomorphic surface context of the artefacts provides additional clues to the temporality of the assemblage. Artefacts on the palaeosol in the centre of the blow-out are argued to be older than artefacts lying on the sandy dune flanks (Lampert and Hughes 1988:147). This is a more detailed approach that those of Smith *et al.* (1991) and Veth *et al.* (1990), yet it only provides a maximum age for surface accumulation, not enabling occupational chronologies to be constructed.

Robins (1993) used a combination of thermoluminescence dates from sediment and radiocarbon determinations from heat retainer hearths to establish a chronological context for surface archaeological deposits in arid southwest Queensland. He (1993:237) predicted that the majority of the surface archaeology on hardpan surfaces would be late Holocene in age. This hypothesis was tested through the excavation and radiocarbon dating of seven hearths from surface contexts and the thermoluminescence dating of two excavated samples from within hardpan deposits. The oldest radiocarbon determination from the surface hearths returned an age of 1730± 80 years BP (SUA 2942). Robins (1993:237-38) concludes that while this does not definitively demonstrate that all the archaeology on the hardpans is late Holocene in age, it suggests that the majority is relatively recent in age. The two thermoluminescence dates from the excavation of two separate hardpan deposits returned ages of 11,700± 1,300 years BP (W999) and 7,900± 500 years BP (W802). He (1993:238) argued that these indicate that the upper units of the hardpan surfaces are terminal Pleistocene/early Holocene in age and that the archaeology within these units is likely to be of a similar age despite the difficulties of obtaining dates in direct association with cultural material. Although this work represents an important attempt at developing an understanding of the temporal character of surface deposits, it is inappropriate to extrapolate this information to the entire study area. The small number of determinations does not really test the hypothesis at all.

In summary, while none of the studies reviewed above have developed detailed occupational chronologies for surface archaeological deposits, they all demonstrate a concern with understanding the temporal character of deflated deposits. In doing this, some of the examples have moved beyond the simple aim of establishing the earliest date for occupation of various regions and have begun to investigate the overall temporal unit represented in surface accumulations.

Establishing an Occupational Chronology for Deflated Surface Deposits at Stud Creek

The Western New South Wales Archaeological Programme (WNSWAP) has demonstrated that not all land surfaces containing archaeological distributions have the same depositional and chronological histories (Fanning and Holdaway 2001a). An archaeological chronology for lagged surface deposits at Stud Creek was established through the radiocarbon dating of 28 heat retainer hearths. This revealed two distinct phases of hearth construction beginning at 2000 BP and separated by a gap at least 200 years in duration (Holdaway *et al.* 2002a). Combined with detailed dating of the sedimentary history of the creek valley, the radiocarbon determinations from the hearths provide the chronological framework for interpreting the occupational history of Stud Creek.

Direct historical links cannot be established between stone artefacts and specific hearths because of the deflated nature of the deposit. As reviewed in Chapter One, the sedimentary history of the valley indicates that the age of the surface upon which the artefacts occur is no older than 5-6,000 years, and is possibly as recent as 2,000 years (Fanning 1999, Fanning and Holdaway 2001a). This provides strong evidence that the majority of stone artefacts were deposited during the last two thousand years (Fanning and Holdaway 2001a). As with the JSN palimpsest deposit from Central Australia, it is not possible to divide the Stud Creek artefacts into separate temporal assemblages.

The identification of a hiatus between two major periods of hearth construction is a significant outcome of the Stud Creek hearth-dating programme. This indicates that occupation is not continuous, highlighting the variable character of occupation during the last 2,000 years (Fanning and Holdaway 2001a). Fanning and Holdaway (2001a:101) argue that in this context it is incorrect to attribute the occupation of Stud Creek to a single function because there is a clear break between periods of hearth construction indicating a change in the occupation of the valley. The factors that account for the hiatus in the Stud Creek chronology are difficult to identify, although sampling error and differential temporal hearth preservation were ruled out on the basis of the large number of hearths excavated and the absence of a relationship between hearth age and spatial distribution. Holdaway *et al.* (2002a) note that the hiatus coincides with the so-called Medieval Warm Period environmental fluctuation recognised in paleoenvironmental data from the Northern Hemisphere. Paleoenvironmental data for late Holocene arid Australia are too sparse and too coarse in resolution to evaluate the regional effects of the Medieval Climatic Anomaly. They argue that the nature of the Stud Creek occupational hiatus will be better understood with the development of detailed occupational chronologies for other parts of the arid zone. Only then can the local scale uniqueness of the Stud Creek occupational history be evaluated.

Unlike studies that have obtained a limited number of radiocarbon determinations from isolated and widely dispersed hearths, the Stud Creek approach demonstrates the potential of hearth dating programmes to investigate history and process in the formation of the surface archaeological record. The study emphasised the diachronic variability of occupation over the last 2,000 years. An occupational chronology for Stud Creek could be established because radiocarbon determinations were obtained for a large number of hearths from the one valley. This permitted the identification of temporal patterns within the occupation history of the valley. The detailed geomorphic study of the Stud Creek valley (see Chapter One) combined with the occupational chronology derived from heat retainer hearths established a temporal package for the investigation of place use history as represented in the reduction and discard of stone artefacts.

Archaeological Time and the Accumulation of the Surface Archaeological Record

The occupational chronology from Stud Creek clearly demonstrates the palimpsest character of some arid zone surface deposits. This necessitates a consideration of the temporal characteristics of deposit accumulation. In particular, issues of continuity and discontinuity in occupation and artefact discard within the context of the chronological resolution of archaeological deposits are important considerations in attempts to identify and define diachronic variability in the archaeological record. Furthermore, when there is no clear continuous historical links between discard events through time, as is the case in deflated deposits, it is more appropriate to investigate variability across space and time within clearly defined chronological units.

Perceptions of Archaeological Time

There are many varied perceptions of time within archaeology and these guide interpretations of the past, particularly those that involve the identification of change in the archaeological record at a range of temporal scales (B. David and Lourandos 1997). For much of the history of archaeology, time has been viewed as linear; operating along a temporal axis in which there is a successive causative relationship between events (Siiriäinen 1991:205). The linear nature of this relationship is used both to generate and explain change. Within these frameworks change is frequently represented as a large-scale event (Shott 1996a:280). Recently, archaeologists have looked beyond single models of change, instead arguing that the processes responsible for the formation of the archaeological record can also operate within a multi-scaled temporal framework. These may be discontinuous from one state to another, and have the ability to generate emergent behaviour arising from multiple domains (McGlade and van Der Leeuw 1997:14). For instance, archaeologists from a range of theoretical perspectives now acknowledge that the one-dimensional nature of linear time is an inappropriate analogue for the nature of change (e.g., Bradley 1991, Fletcher 1992, Gosden 1994, Hodder 1987, Shanks and Tilley 1987, M. E. Smith 1992). Instead they emphasise that change occurs at a range of different temporal scales.

Braudel (1980) identified four hierarchical temporal scales, with social time divided into the intermediate and long-term. These are:

- The *longue durée* – the equivalent of geological time, operating over the long-term at the scale of environmental change
- Social time (divided into intermediate term and long-term) – the history of particular groups of people
- Individual time – the equivalent of the event

Braudel's temporal scales are widely used to discuss the temporality of the archaeological record, particularly in terms of short-term and long-term processes, but not all archaeologists agree with the interpretations given to the different scales (Bailey 1983, Duke 1996, Fletcher 1992, Shanks and Tilley 1987, M. E. Smith 1992). All temporal scales contribute to the formation of archaeological deposits, but some argue that the *longue durée* is closest to the temporal character of certain types of archaeological record (e.g., Hodder 1987, McGlade and van Der Leeuw 1997, Stern 1993). Examples from the archaeological record include the long time span represented in the accumulation of Pleistocene deposits in East Africa (Blumenschine and Masao 1991, Stern 1993), and the late Pleistocene cave deposits in southwest Tasmania (Cosgrove and Allen 1996). There are archaeological exceptions, but these are rare. Bailey (1983), for instance, incorporated Braudel's scheme into a model of archaeological time, in which two contrasting notions of time are identified. In the first, the past is explained in terms of the present, reconstructed as a series of present moments. In the second, the past is explained in terms of the past. This is undertaken from the perspective that large-scale historical processes not visible to the individual observer structure the archaeological record over the long-term. For Bailey, these are not mutually exclusive entities, but form an area of disagreement between processual and post-processual approaches, particularly concerning the role of short-term events in the long-term structure of the archaeological record (Hodder 1987:6).

Bailey (1983) relates temporal scale to process, arguing that the short-term is characteristically associated with social processes, while the long-term is more often perceived as a consequence of environmental and biological processes. Some archaeologists (e.g., Bradley 1991, McGlade 1999, Shanks and Tilley 1987) question the association between the long-term and environmental processes. Shanks and Tilley (1987) are sceptical of the definition of long-term and short-term, correctly pointing out that these are not necessarily simple references to chronology. Furthermore, they (1987) argue that Bailey emphasised the importance of environmental processes operating over the long-term, thus under estimating the role of social processes and individuals in material culture change.

Others (Bradley 1991, B. David and Lourandos 1997, Gosden and Lock 1998, Hodder 1987, McBryde 1984, Preucel and Hodder 1996) have suggested that social processes and structures may operate over the long-term. Bailey (1983) acknowledges this, but is critical of such arguments because these have rarely been articulated beyond the scale of the repetition of short-term processes through time. Hodder (1987) and others (e.g., Gosden and Lock 1998) argue that long-term continuity in social, economic and symbolic structures can be identified in the archaeological record. As noted by M. E. Smith (1992) this is a weakness of many, but not all, post-processual interpretations of social processes when these are derived from the substantive uniformitarian projection of ethnography onto the archaeological record. A number of Australian studies (e.g., B. David and Wilson 1999, Hemming *et al.* 2000, Taçon 1991, 1994) illustrate the substantive use of models derived from ethnography without a consideration of the temporality of deposit formation. The reiteration of short-term social processes through time reduces the past to being like the present, and ignores the possibility of identifying historical variability in archaeological deposits. Social interpretations that fail to acknowledge the dynamic temporality of the formation of archaeological deposits through the application of analogy are little more than the projection of the present onto the past (Yoffee and Sherratt 1993:8).

In many cases, the ability to identify short-term processes, particularly the relationship between social time and material culture change is constrained by scale (Preucel and Hodder 1996), particularly the temporality at which the archaeological record can be investigated. David and Lourandos (1997:10) view this as a limitation because it means that it is difficult, if not impossible, to represent human behaviour in a meaningful way at the scale of the individual. Others (e.g., Stern 1994) disagree, instead arguing that the investigation of long-term patterns in the archaeological record offers greater potential for investigating the relationship between the composition of the archaeological record and human behaviour.

The difficulty of detecting short-term processes in deflated arid zone surface distributions does not denigrate the record as a source of information or deny the importance of social processes in the formation of archaeological deposits. Deflated deposits, by their nature cannot be teased apart into a series of short-term historical processes. For the types of questions addressed in the present study it is not desirable to attempt to deconstruct the record into a series of short-term processes because this would obscure the very patterns discernible in the history of place use. Viewing the archaeological record as a dynamic construction reflecting both behaviour and geomorphic processes operating over varied amounts of time need not and cannot rule out the significance of social processes in the formation of accumulated deposits.

Historical information may inform on the particular social processes associated with the formation of a specific archaeological deposit at one point in time, but this cannot be uncritically applied across space or through time. As with adaptive strategies, social processes are dynamic, with change occurring at multiple temporal and spatial scales. Therefore, social processes are not essential categories. Subsequent groups may view the material remains of past occupations in a very different way to the occupants who originally deposited the material. They may see it as rubbish or a resource for further reuse and modification (Camilli and Ebert 1992). The same can be said of other archaeological manifestations. Fletcher (1995) for example notes that walls and other features control how space is used and this in turn influences the long-term structure of the archaeological record. Physical landscape features also influence the use of space (David and Wilson 1999).

This review does not argue that environmental processes are more important than social processes in the formation of the western New South Wales archaeological record or vice versa. Instead, synchronic notions whether they are environmental or social have been challenged. Interpretation of the archaeological record at the scale of the long-term is promoted because this best reflects the temporal scale over which the accumulated surface deposits in western New South Wales have formed. It is neither desirable, nor possible at present, to dissect the record within these into a series of short-term slices of time. Deposits with multiple radiocarbon determinations indicate that the composition of these assemblages represents pattern accumulated over the long-term. It is this accumulated pattern, rather than many individual short-term discard events that permit an investigation of place use histories across different locations. The point is to look at the record at a series of scales and as the outcome of historic processes.

Time Resolution and the Palimpsest Problem

Time resolution in stratified and non-stratified deposits is as important an issue in Australian archaeology as in other parts of the world, particularly in discussions of change through time (Frankel 1988). A broad temporal scale of investigation can mask minor variability by collapsing all "...variation into one average form of behaviour..." (Frankel 1993:25). Frankel (1993:25) also notes that the fine temporal scale required for the identification of trends and disconformities is rarely available in the Australian archaeological record (but see Holdaway *et al.* 2002a). In considering the temporal nature of the archaeological record, some archaeologists have been reluctant to abandon the pursuit of the elusive single event. Mardaga-Campbell (1986) argued that the failure of Australian archaeologists to identify living floors reflects the practice of excavating sites with arbitrary spits rather than using the characteristics of archaeological deposits as a guide. In contrast, others (e.g., Villa 1982) note that the time resolution of stratified deposits is often too coarse for the identification of single events, a fact acknowledged by many Australian archaeologists (Cosgrove *et al.* 1990, Cosgrove and Allen 1996, Holdaway and Porch 1996, Holdaway *et al.* 1998, E. Williams 1998). Despite this recognition, Australian archaeologists have struggled to develop frameworks that take into account temporal resolution and the structure of assemblage accumulation.

Murray (1993) challenged the notion of the "palimpsest problem" by arguing that the archaeological record should be perceived as a singular record of human behaviour. He notes that many contemporary problems regarding time and the nature of the archaeological record can be linked to the theoretical challenge of the recognition of deep time faced by 19th Century prehistorians. Their solution was to rationalise the past by interpreting it in terms of the ethnographic present; thus creating a prehistoric past based on ethnography. This has created problems for modern archaeologists who acknowledge the conflicting temporality between the ethnographic and archaeological records and the relative scale at which behavioural variability can be identified (Wobst 1978). This results in a "crisis of interpretation" for archaeology, more specifically between the nature of the record (deep time) and the perception that this record represents a distinct order of reality (Murray 1993:178). To counter this, many archaeologists have sought finer temporal scales of resolution. However, the application of ethnographic parallels to systems with a great temporal depth is problematic and will not permit the identification of the material correlates of ethnographic scale behavioural events in the archaeological record (Murray 1993, M. E. Smith 1992). Finer temporal scales of resolution will not necessarily solve this problem because even these will only rarely permit the identification of single events. Despite this, appropriate explanatory frameworks that go beyond ethnographic scales of interpretation and consider the temporal characteristics of archaeological data have rarely been developed (Fletcher 1992, Frankel 1988, 1993).

Stern (1993, 1994, 1995) discusses the inability of archaeologists to deal with the temporal nature of the archaeological record. Like Murray (1993), she argues that the palimpsest nature of archaeological deposits means that they are not amenable to ethnographic temporal scales of interpretation. Instead, the goal of archaeology should be to explain patterning in the archaeological record that is the result of long-term processes (Stern 1994:89). As Stern (1994) notes, archaeological deposits are often time-averaged records of both cultural and non-cultural processes. Time averaging is not an average over time; the sum number of artefacts cannot be divided by the total amount of time represented in a deposit, nor is it assumed that a stable set of behaviours resulted in the accumulation of palimpsest deposits. Moreover, unlike other methods of studying the temporality of accumulation, for example age-depth curves, widely used by Australian archaeologists to model discard rates through time (e.g., David *et al.* 1990:81-82, Veth 1993:46, 55, 64, 66) time averaging does not assume that materials accumulated at a constant rate (Stern 1994:102). In contrast, the time averaged framework refers to the structure of aggregated material from multiple episodes of behaviour. Time averaging is founded on the basic stratigraphic concept that the accumulation of sediment and in-turn artefacts is frequently intermittent and localised in certain contexts (Stern 1994:90). Therefore, periods of accumulation may be separated by periods of non-accumulation. Acknowledgement of this requires the development of interpretative frameworks that explain patterning visible in palimpsest assemblages without recourse to ethnographic scales of explanation (Edwards 1995, Stern 1994).

Time averaging is based upon the principle of the minimum stratigraphic unit, the smallest sedimentary package available for the study of archaeological distributions (Stern 1994). In East Africa, Stern (1993) used the concept to investigate the distribution of archaeological materials in multiple geological beds. Time averaging has often been discussed in relation to stratified deposits (Holdaway 2000, Stern 1993, 1994). Recently, Holdaway *et al.* (2000) and Holdaway and Fanning (2001a) have argued that eroded land surfaces with distributions of stone artefacts and heat retainer hearths in western New South Wales represent time averaged deposits. The age of the surface upon which artefacts are distributed provides a maximum date for the commencement of artefact discard, thus defining the equivalent of the minimum stratigraphic unit.

Discard and the Accumulation of Archaeological Deposits

Materials that have accumulated in the archaeological record can be interpreted within many different temporal and spatial frameworks (Fletcher 1995, Olivier 1999:117). To understand the scope of these different frameworks requires a consideration of artefact discard histories.

The time-averaged nature of the record raises issues regarding the temporality of behavioural and ecological processes. Stern (1994) notes that these processes operate over different time spans, and that these are far shorter in duration than the accumulation of the archaeological record. The duration of these processes may also change between subsequent occupations. Not all locations may be used the same way throughout time and the amount of time spent at a location can vary between occupations. Environmental variability (Fanning and Holdaway 2001a), social factors, population dynamics (Binford 1980) and discontinuity in occupation (Holdaway *et al.* 2002a, Shennan 2000) may change the character of occupation and artefact discard. For instance, population fluctuation may alter the use of space. An area of occupation may expand and recede according to changes in population; therefore artefacts may no longer be discarded in certain areas.

Rather than deconstructing the record into separate functional events, a time-averaged perspective permits an investigation of the amount of behaviour performed at different locations through a consideration of artefact discard. Binford (1980) argued that occupation in environments with a limited distribution of critical resources, e.g., water would be restricted to a few select areas, increasing the year to year redundancy in the use of certain locations and resulting in the greater potential accumulation of archaeological materials. The dynamic nature of environment variability through time can also act as a counter to redundancy. As suggested by Holdaway *et al.* (2002a), environmental oscillation may have a bearing on the continuity of place use. Redundancy is a time transgressive variable with its own temporal trajectory. In many situations, redundancy may not necessarily operate over the same time span as that represented in the accumulation of the archaeological record. As landscape features change, e.g., a waterhole may disappear, leading to a change the level of redundancy in place use. In this situation, people may no longer visit this location and artefact discard will cease or the duration and intensity of occupations may change resulting in the discard of different proportions of artefacts.

Change, Variation and the Temporal Character of the Record

There is a dichotomy between dynamic cultural processes and the static archaeological record (see Binford and Sabloff 1982:137). In Australia, change is often framed as a continental wide phenomenon, for example in models of late Holocene intensification (Lourandos 1983, 1985, Veth 1993). This largely reflects the way archaeologists have approached the record rather than the character of the record itself (Holdaway 1995a). The surface archaeological record as a palimpsest does not rule out investigating change. To investigate change necessitates that the concept be reconsidered as diachronic variability in records where there are no clear historical relationships within and between assemblages (see Dibble 1991:40-41).

Binford (1982) discussed the concept of change in archaeological deposits that represent long-term accumulation. He notes that archaeologists often identify chronological variation in the density of artefacts from individual sites. This variation is commonly interpreted to represent significant organisational changes in cultural systems i.e., culture change. Binford (1982) disagrees with this interpretation, instead arguing that in many instances variation represents change in the positioning of the system state rather than the decline of one system and the appearance of another. This does not imply that Binford views the system state as a synchronic organisational framework. Indeed, his (2001:51) notion of the system state as any "dynamic system (that) may exhibit significantly different properties when observed at different times" emphasises change. For Binford change is driven from within systems and variation in the archaeological record is not the product or signature of different cultural (ethnic) groups (see Binford and Sabloff 1982).

Braudel's notion of the relationship between the long-term and change is criticised by a number of archaeologists (e.g., McGlade 1999, M. E. Smith 1992). M. E. Smith (1992) argues that Braudel's concept of the *longue durée* as a structural, essentially static rather than dynamic entity creates problems for archaeologists interested in change, particularly over the long-term. Instead, he suggests that archaeology requires a longer temporal perspective than the *longue durée* for the investigation of change. McGlade (1999) regards the *longue durée* as an environmentally deterministic concept, arguing that it swamps the role of contingent events in the structure of the archaeological record, replacing it instead with a monolithic perception of structure as a slowly changing entity. Additionally, he (1999) states that Braudel's model provides no provision for episodic or abrupt change, thus emphasising continuity at the expense of discontinuity. McGlade (1999) concludes that this results in a lack of attention to how different temporalities subsume a variety of differential rates of change in terms of economic, social and political criteria.

Alternatively, McGlade (1999) advocates a non-linear perception of time that emphasises the role of necessity and chance as the major agents generating change. Following Shermer (1995) this is organised within the framework of contingent-necessity in which the occurrence of events at the same time results in a specific type of action by constraining prior conditions (McGlade 1999:157). Necessity takes a historical system down a specific path until a point is met where multiple paths become available (McGlade 1999). Contingency is the major driving force behind change; at the point of multiple paths when it selects a new path for the system, which over time develops its own necessity (McGlade 1999). This non-linear view of time is not compatible with perceptions of history as a developmental sequence or narrative (McGlade 1999:147). Instead, history is continuous and discontinuous and the product of contingent and deterministic factors (McGlade 1999:159).

Duke (1996) argues that Braudel's scheme is useful for investigating the relationship between short-term events and long-term structures. He (1996) proposes a structure-event model in which events constitute a marker of transition and a point of analysis of changing social structure organisation. In a discussion of projectile points in North American Plains assemblages, he regards procurement and processing tools as long-term structures representing both economic behaviour and social symbols of human action and social negotiation. He argues that the intense investment of time and effort in making projectile points constitutes a long-term structure. Stylistic change over this period represents notions of power and prestige associated with hunting, in particular rapid stylistic change can be linked to male perceptions of status. Duke (1996) suggests that hunting, as symbolic action, is a process by which punctuated change occurs in long-term structures. While it is conceivable that hunting is a long-term structure, Duke's model is dependent upon synchronic meaning derived from analogy. For the model to work, it must assume that the ideology of hunting remained static across many multiple generations.

An alternative to identifying and explaining change comes in the form of behavioural ecological models. A behavioural ecological model for a series of Pleistocene caves in southwest Tasmania has been proposed (Cosgrove 1999, Cosgrove and Allen 2001, Cosgrove *et al.* 1990). Although derived from stratified deposits, faunal and lithic assemblages from the caves are acknowledged to represent palimpsests at the scale of decades and perhaps even greater periods of time (Cosgrove and Allen 1996:23). The model argues that in the southwest of Tasmania, where resources are clumped and predictable, people moved between resource patches within their estate ranges (Cosgrove 1999:394). Movement between these patches was probably not continuous and may have involved seasonal contact with outside areas (Cosgrove 1999:394). This model is referred to as the 'Southwest System' (Cosgrove and Allen 2001:422). However, within the model of resource exploitation there is evidence for temporal and spatial variation in the composition of faunal and lithic assemblages from southwest Tasmania (see Holdaway 2000). Cosgrove and Allen (2001) suggest that temporal variation in the faunal assemblages may indicate different land use patterns either side of the Last Glacial Maximum. The chronological record indicates long discontinuities of occupation (Holdaway and Porch 1996). Therefore, the caves did not form part of the same settlement system, but instead represent discontinuous use and accumulation beyond ethnographic scale land use models. Temporal and spatial variability in discard and assemblage accumulation indicate that the notion of a 'Southwest System' enduring over a period of 20,000 years is unlikely. The temporal and spatial differences in southwest Tasmanian assemblages represent the variable nature of the occupation histories of places over the long-term. Within these variable occupation histories, there are certain long-term structures that result in similar adaptive strategies and similarities in the material culture, such as the exploitation of similar lithic sources, but the effect of occupational hiatuses on assemblage formation indicate that the record does not represent a single behavioural system.

To investigate change requires the definition of multiple comparative units. The distinction between continuous and discontinuous change is important. If change is a continuous variable, it is difficult to define comparative units because elements are in a constant state of metamorphosis. A problem with some approaches is that the identification of change in the archaeological record is predicated on the notion that artefact discard is continuous. Of course this is not always the case, especially in an archaeological record created by hunter-gathers where discard at individual locations is discontinuous and separated by blocks of time between episodes of discard (Cosgrove and Allen 1996, Holdaway *et al.* 2002a, Shennan 2000).

Archaeologists have used multiple blocks of time to investigate change through time in single archaeological deposits. This is based upon an assumption of continuity and the existence of historical linkages between artefacts in different temporal units (Dunnell 1980). In some cases this change can be identified in artefact seriation, but this is based on essentialist typological categories (O'Brien and Holland 1995). In some assemblages there is no direct historical relationship between the former and the latter. The same can apply for analyses between multiple locations. For instance, the assemblages from southwest Tasmania represent a general set of behaviours that resulted in the discard of similar assemblages. A recent analysis of the Bone Cave assemblage from southwest Tasmania (Holdaway 2000) indicates temporal variability in local lithic raw material reduction. Similar types of raw material were exploited throughout the occupation of the cave, but the degree to which these were reduced changed. This is an example of the type of temporal variability identifiable in archaeological deposits that have accumulated over long periods of time that may not have clear cultural relationships. At a very broad temporal and geographic scale these assemblages can be viewed as a single cultural system. This in part reflects the resolution of the record (Cosgrove and Allen 1996).

Where archaeological deposits have accumulated through discontinuous occupation and discard, variability rather than change is evident in the record. The notion of variability removes the requirement for the high-resolution linkages required to identify change. Contra the approach of Veth (1993), archaeologists cannot study deflated surface records at the scale of a single settlement system. The long-term nature of surface accumulation indicates that artefact discard may occur under different settlement strategies, particularly in areas where significant environmental change has taken place. In the absence of these linkages, the challenge for archaeologists is to identify the appropriate investigative scales at which variation can be identified (see Frankel 1993). These scales are both temporal and spatial. A series of spatially separate surfaces that represent the accumulation of materials within a common block of time can be investigated to study spatial variation in place use histories and artefact discard at a landscape scale. Alternatively, spatially separate surfaces that represent artefact accumulation across different temporal contexts can be compared. These will permit considerations of land use over the long-term, but not in the sense of conventional settlement pattern models. The point is to consider what types of behavioural strategies might be compatible with the temporal scale that the archaeological record can be investigated. As O'Brien and Holland (1995) suggest the focus of archaeological research should be on variability in the record and the dynamic processes that lead to this.

Approaches to the Spatial Analysis of Surface Artefact Distributions

Investigations of 'site structure' as represented in the spatial distribution of artefact discard provide a method for articulating spatial patterns with the temporal character of occupation history, particularly in terms of the intensity and duration of occupation (Wandsnider 1996). Spatial analysis has also emerged as an important investigative tool in surface archaeology, especially for evaluating the degree to which geomorphic processes have influenced artefact distribution (Fanning and Holdaway 2001b, Hawkins 1998, Kvamme 1998).

An associated issue arising from the spatial analysis of extensive surface distributions is the definition of landscape and the selection of appropriate sampling units. These are complex concepts of which there is no single or correct definition. The definition of landscape and the range and scale of sampling units depends on the nature of the archaeological record being investigated and the questions posed. Methodological uniformitarian assumptions drawn from Australian arid zone ethno-archaeological studies (O'Connell 1987) can provide a guide, but not a direct analogue. O'Connell's study from Central Australia suggests that areas of at least 10,000 square metres are required for the investigation of spatial patterning in artefact distributions. This scale is many orders of magnitude greater than most conventional investigations of spatial patterning in Australian surface archaeology.

Occupation Intensity, Occupation Duration and Spatial Structure

Increased artefact density has often been associated with greater intensity of occupation. Dense artefact distributions are interpreted to represent the most intensively occupied places on the landscape. At a basic level this is a logical expectation, but several factors must be considered before this interpretation is accepted, the most important of which is the length of the temporal unit represented in time averaged deposits. Other factors include the geomorphological history of the place and the archaeological methods applied.

Spatial Scales and Sampling Units

Clarke (1977) defined three scales at which spatial analysis can be undertaken:

1. Micro level – within structures, e.g., a shelter, house, room and other features
2. Semi-micro level – within a site, e.g., the distribution of artefacts within a location
3. Macro level – between different sites

A problem with Clarke's hierarchy of spatial analysis is that it is founded upon the concept of the site. Archaeologists who favour landscape, distributional and non-site scales of analysis have challenged the reality of the site as an empirical unit. In hunter-gather archaeology considerable ambiguity exists as to what actually constitutes a site. A common definition is that a site is a place that was a major locus of activity (Vinnicombe 1984). By implication dense distributions of artefacts are perceived to represent significant amounts of behaviour and are attributed the status of site at the expense of lesser artefact densities. Further complicating the application of site based studies is near continuous distribution of artefacts visible in regions with good ground exposure. Artefact distributions lack clear boundaries for the definition of sites and assemblages (Holdaway *et al.* 1998). Nonsite archaeology (Foley 1981, Thomas 1975) developed in response to these factors. An alternative is to investigate spatial patterning that is the result of long-term behaviour (Zvelebil *et al.* 1992). In addressing this issue, Ebert (1992:189) states:

> the information available to the archaeologist through the analysis of the archaeological record is a time exposure of the structure of human reuse of the landscape. One way this structure can be analysed and measured is by examining the spatial scales of repetition of the patterning of artifacts and their characteristics.

Landscape scale studies have developed as a method of investigating patterning in artefact distribution over extensive spatial areas. Spatial studies (e.g., O'Connell 1987, Yellen 1977) suggest that investigative units need to be many times larger than that of most conventional archaeological excavations.

The scales at which spatial analysis can be undertaken in excavated deposits are limited (Ebert 1992). Unlike open surface distributions, those within caves and rockshelters are constrained by walls (Holdaway *et al.* 1998). The large scale over which spatial distributions may be recorded for surface deposits is advantageous because multiple scales of spatial analysis can be undertaken. A complication is that it is difficult to define the appropriate size of sampling units. Indeed, it is impossible to know the size and shape sampling units to employ without some prior knowledge of the distribution to be sampled (Wobst 1983, Zvelebil *et al.* 1992). Ethnoarchaeological research (O'Connell 1987, Yellen 1977) suggests that the scale of spatial patterning will vary between distributions.

It is impossible to define a universal sampling unit. In some contexts poor conditions of visibility may limit artefact exposure, thus requiring larger spatial areas to be investigated to both identify pattern and obtain a sufficient sample of artefacts for analysis. Unit size cannot be equated with an ideal sample proportion of a distribution. An alternative is to consider the nature of artefact discard at repeatedly used places. In some cases, particularly at places without constraining features, repeated occupation through time may result in the discard of artefacts over large areas, and there is a low probability that artefacts will be discarded in exactly the same location as those of previous occupations. Although there is likely to be some redundancy in the use of space in relation to topographic features (Binford 1980) archaeological features can also regulate the use of space in the opposite way. For instance, Fletcher (1995) argues that material i.e., walls, features and physical structures over the long periods of time have a distinct regularity and restrictive function in managing community life. Thus material is an important component that regulates how people utilise space. Hearths can be thought of as a regulator of the use of space in hunter-gather societies. Wandsnider (1992) also notes that reuse and reoccupation of a location may not necessarily result in the same use of space with previous occupations. For instance, during the subsequent occupation of a place there are many options for how people interact with hearths from a previous occupation. They might reuse the hearths, construct new hearths next to the old hearths or construct new hearths in areas away from old hearths. These decisions will influence the spatial discard of artefacts.

Landscape

Landscape is a ubiquitous term, the use of which is not restricted to archaeology (Gosden and Head 1994, Knapp and Ashmore 1999). Within archaeology there are a multitude of different definitions of landscape (Anschuetz *et al.* 2001). Some archaeologists emphasise social landscape, arguing that experience of place participates in the construction of identity (e.g., David and Wilson 1999, Gosden and Lock 1998, Taçon 1994) while others (e.g., Waters and Kuehn 1996) emphasise that archaeological landscapes are both cultural and geological. The definition of social landscapes in the past often depends on the substantive projection of ethnography onto the archaeological record (e.g., B. David and Wilson 1999, Taçon 1994). Social and cultural factors influence the way people utilise landscapes, however, archaeological interpretations of social landscapes rarely go beyond assertions that co-location in space between various artefacts and features is evidence of social factors influencing artefact discard.

Wandsnider (1998b:96) notes "archaeological landscapes are both cultural and geological phenomena". Geological landscapes are not static, since geomorphic processes are constantly operating at different temporal and spatial scales (Wandsnider 1998b:96). For example, the accelerated exposure of the surface archaeological record of western New South Wales during the last 150 years (Fanning 1999), although occurring episodically, is an example of how landscapes demonstrate change at different spatial and temporal scales. Nor are archaeological landscapes static in the sense that they are not the product of one single cultural system (Wandsnider 1998b:96). Zvelebil *et al.* (1992:194) provide a definition of an archaeological landscape as "a past surface within a defined span of time, which is subject to antecedent features and successive modifications". They (1992) emphasise the dynamic nature of the landscape and clearly regard the archaeological landscape as a palimpsest. These palimpsest landscapes retain material that is the remnant of multiple settlement systems, and these rarely, if ever, retain all the components of an individual system. A combination of these definitions is useful for describing the deflated archaeological landscapes of western New South Wales.

Summary

This chapter argues for the study of the surface archaeological record at multiple temporal and spatial scales. In particular, it has been emphasised that surface deposits particularly those of western New South Wales are time-averaged palimpsests that may be beyond ethnographic scales of interpretation. From this perspective, synchronic interpretative frameworks have been questioned. These imply that Aboriginal people were an unchanging people in an unchanging landscape (e.g., Cane 1984:274), a view passionately, although largely unsuccessfully, challenged by many Australian archaeologists. Instead, it is argued that a diachronic perception of function is desirable, one that acknowledges that the function of artefacts, places and landscapes is multidimensional and may change according to different contextual circumstances.

The palimpsest character of the record suggests that it is appropriate to investigate long-term accumulated patterns, rather than seek to dis-entangle the record into a series of short-term events that cannot be linked to occupational chronologies. The different temporalities that are represented in the surface archaeological record illustrate the multitemporal nature of archaeological assemblages. This does not deny that short-term processes also contribute to the accumulation of archaeological records, but palimpsest surface records of arid Australia may rarely document accumulation at such fine temporal scales. Scales of interpretation and analysis need to be matched.

The desirable outcome of the approach outlined here is the development of a history of landscape use, which has eluded many archaeologists who approached landscape from a processual perspective. This does not detract from our ability to investigate the history of different places. Models derived from ethnography and attempts to obtain the finest chronological resolution possible are not appropriate or necessary for the writing of these histories. Investigating long-term place use at a range of scales and with an understanding of the temporality of the record provides a profitable avenue of research for arid zone surface archaeological studies.

Chapter Three: Studying Stone Artefact Assemblage Composition

Introduction

Stone artefacts are the most durable component of the Australian archaeological record. They are widely distributed and are found in both stratified and non-stratified contexts. Their ubiquity has resulted in a long history of research, which has often been undertaken at local, rather than regional scales. As a consequence, detailed knowledge of stone artefact assemblages has not been obtained for many parts of Australia. This has inhibited efforts to identify both spatial and temporal changes in artefact form and variation in assemblage composition.

The demonstration of change in stone artefact assemblages is a common aim of many Australian lithic studies, particularly identifying the chronological context for the appearance of different elements of the Australian small tool tradition. Since the early 1980s, many attempts to identify spatial and temporal change have been linked to various models of late Holocene intensification. Apparent increases in the rate of artefact discard during the last 4,000 years are argued by some to represent evidence for population increase and the development of more complex economic and social relations (Lournados 1983). In general, studies of stone artefact assemblage composition have not been undertaken within an analytical framework that permits the investigation of these issues. This chapter develops an economic approach to the study of stone artefact assemblage composition that facilitates an investigation of the relationship between raw material utilisation and artefact discard over the long-term.

Australian Stone Artefact Assemblages

Australian stone artefact assemblages have been broadly divided into two distinct temporal groups. The first, the Australian core tool and scraper tradition (Bowler *et al.* 1970) dates from earliest occupation during the Pleistocene until the mid-Holocene. The second, the Australian small tool tradition (Gould 1969) is widely thought to have appeared during the mid-Holocene and persisted into the contact period i.e. the last 200 years. The appearance of the Australian small tool tradition has been argued to be the only major change occurring in assemblage composition (but see Bird and Frankel 1991a for critique of this argument). This view has been challenged on the grounds that the two assemblage groups are not as homogeneous as originally perceived, with elements of each appearing and disappearing at different times (Hallam 1985, Hiscock and Attenbrow 1998, R. Jones 1985).

The Core Tool and Scraper Tradition

The core tool and scraper tradition is a term first used by Bowler *et al.* (1970) to describe the Pleistocene assemblages from Lake Mungo in western New South Wales. They (1970:52) argued that the main components of the assemblages (core tools, steep edge scrapers and flat scrapers) were generally characteristic of older Australian assemblages that persisted from 30,000 years ago until the introduction of a suite of new specialised forms at approximately 6,000 BP. As a result, the term has been widely adopted as a description of assemblages from the Pleistocene to the Early Holocene. Subsequent research has confirmed the assertion of Bowler *et al.* (1970) that there is considerable spatial and temporal variation in Pleistocene stone artefact assemblages. For example Cosgrove (1995) reported the presence of small finely retouched thumbnail scrapers from Pleistocene deposits in Southwest Tasmania. The Tasmanian thumbnail scrapers contrast to the large steep edge scrapers discovered at Lake Mungo. Based on a re-analysis of the Lake Mungo assemblages, Hiscock and Allen (2000) argue that significant regional variation between Pleistocene assemblages challenges concepts of temporal and spatial uniformity.

A particular area of interest in Australian studies of pre mid to late Holocene stone artefact assemblages is the apparent lack of change in assemblage composition during the Pleistocene–Holocene transition (Holdaway 1995a). The transition is a period of considerable global environmental change marked by significant changes in the composition of assemblages from other areas of the world (Meltzer 1988). Change in the frequency of artefacts and the use of raw materials has been identified in Australian Pleistocene assemblages, however the identification of large-scale change in assemblage composition during the transition has eluded archaeologists. Holdaway (1995a) suggests that this unique pattern may have more to do with the methods employed by Australian archaeologists rather than the nature of stone artefact assemblages. He argues that they have rarely adopted the regional scales required for the identification of change and variation in assemblages dating to the transition.

The Australian Small Tool Tradition

The term 'Australian small tool tradition' was first proposed by Gould (1969) to describe the appearance of a new suite of tool forms including, backed blades, tula and burren adzes and points (both unifacial and bifacial), during the mid to late Holocene. The standardised form of these new types has been argued by some to represent functionally specific elements of specialised hafted toolkits (Mulvaney 1969). The concept was rapidly adopted by Australian archaeologists, with the new artefact forms serving as temporal markers for the definition of mid to late Holocene assemblages (Lourandos 1997:287). Numerous studies examining the distribution of these components at a continental scale were undertaken, often attempting to define type boundaries (Pearce 1974). The results of these studies suggest that different components of the Australian small tool tradition had distinct distributions. Early studies argued that backed blades were confined to the southern two-thirds of Australia, adzes (tula and burren) to arid and semi-arid regions and points (bifacial) to the north. Subsequent research has challenged these interpretations, instead arguing for more widespread and overlapping distributions (Davidson 1983, Hiscock and Hughes 1980, McNiven 1993, Mulvaney 1985).

More recently, attention has turned to the chronological context of the Australian small tool tradition, with some archaeologists (e.g., Hiscock and Attenbrow 1998) arguing that some new artefact forms appear before the mid Holocene. Hiscock and Attenbrow argue for the presence of a small number of backed artefacts[1] dating to the early Holocene in the Mussel and Loggers Shelter sites at Mangrove Creek in eastern New South Wales. McNiven (2000) unconvincingly developed this argument further by suggesting that backing retouch is present on a number of artefacts, including thumbnail scrapers from the Pleistocene deposits of Bone Cave in Tasmania. He states that this is particularly significant considering that the Australian small tool tradition is argued to not have been present in Tasmania. Jones and Johnson (1985) have also argued for an early date for the appearance of point technology based on the presence of one broken bifacial point in deposits dated to 5,700 years ago from the Nauwalabila I site in the Northern Territory.

There are two major problems with these arguments. Firstly, in each case samples are limited and it is unclear whether or not the retouch is a by-product of tool maintenance. Secondly, the assemblages from whom these examples are drawn lack evidence for the wider range of technological organisation associated with the production of backed artefacts during the mid to late Holocene, especially specialised core reduction strategies (see Hiscock 1993). Therefore, it is unlikely that these isolated backed artefacts represent type fossils with clear historical associations to more recent examples.

Models that advocate a mid to late Holocene date for the Australian small tool tradition view the appearance of new artefact types and the arrival of the dingo at approximately the same time as more than coincidental (Bowdler and O'Connor 1991). Therefore, the Australian small tool tradition was an external development, brought to Australia by a recent group of migrants. In contrast, those who argue for early development (Hiscock and Attenbrow 1998, McNiven 2000) perceive the Australian small tool tradition as an indigenous invention, resulting from a long technological tradition of backed retouch. Others (Mulvaney 1985) are cautious of accepting a link between the Australian small tool tradition and the dingo, arguing that the appearance of backed blades precedes the arrival of the dingo. Backed blades, pirri points, tula and burren adzes, and thumbnail scrapers are components of the Australian small tool tradition found in western New South Wales assemblages.

Typology and the Chronological Context of Surface Deposits

Throughout the 1970's and 80's the core tool and scraper tradition and the Australian small tool tradition were used as temporal markers to establish chronological contexts for assemblages derived from surface deposits (Davidson 1983, Ross 1981, 1984). A notable example is Ross's research in the semi-arid Mallee region of northwest Victoria. Ross (1981,1984) argues that the absence of artefacts characteristic of the Australian small tool tradition in northern Mallee assemblages indicates that these pre-date the late Holocene. In support, Ross (1981) cites a radiocarbon determination of 7650±110 years BP (SUA-722) from shell associated with a hearth and lithic scatter. Assemblages from the southern Mallee lack large flake tools and are instead dominated by small artefacts associated with the Australian small tool tradition. Two radiocarbon determinations on shell of 2310±80 years BP (SUA-109) and 1470±80 years BP (SUA-110) are argued to support the interpretation that the southern Mallee assemblages are no older than 4,500 BP (Ross 1981).

A number of problems have been acknowledged with relying on typology and a limited number of radiocarbon determinations to establish chronological context (Smith and Clark 1993). Stone tool typology is a poor chronological marker in surface deposits (Bird and Frankel 1991b:185). Even if the Australian small tool tradition is unequivocally assigned to the mid to late Holocene, the time transgressive nature of assemblage accumulation complicates the use of artefact forms as definitive chronological markers. As argued in Chapter One, aggregated deposits on lagged surfaces may comprise artefacts accumulated over long periods of time. For example, the JSN assemblage from the Strzelecki Desert represents the potential accumulation of artefacts over 10,000 years (Smith *et al.* 1991). In these contexts, backed blades and other items traditionally assigned to the late Holocene may indicate that discard occurred during a specific temporal period, but other artefacts may have been discarded during periods preceding this. Inferring temporal context on the basis of the presence or absence of different artefact forms alone is problematic.

Another approach is to investigate the nature of reduction strategies as reflected in debitage. Hiscock (1986) argues that the characteristics of stratified assemblages can be used to seriate surface assemblages in other areas of the New South Wales Hunter Valley. Although a more comprehensive consideration of temporal context than that derived from tool typology alone, this approach faces similar problems to typological approaches when confronting deflated palimpsest assemblages. Surface assemblages may represent accumulation over long periods of time, and artefacts are difficult to seriate because relative chronological relationships have been lost. In addition, while some broad temporal characteristics may be identified, it will be impossible to assign the majority of artefacts to a temporal period because of the general nature of core reduction. Artefacts within several metres of one another may have been discarded several thousand years apart.

Temporal context in surface deposits can be understood through the concept of the minimum stratigraphic unit. The surface upon which artefacts are distributed provides a maximum age for the temporal unit in which artefact discard occurred. While it might be possible to estimate this following more comprehensive studies of overall assemblage composition (e.g., Hiscock 1986) the temporal nature of artefact accumulation in surface assemblages is best understood through the establishment of absolute occupational chronologies within geomorphic contexts.

Themes in the Interpretation of Australian Stone Artefact Assemblages

Several themes have dominated the interpretation of Australian stone artefacts. Early studies focused on the establishment of typological sequences that would permit the seriation of artefacts within cultural-historical frameworks. Typological approaches were replaced during the 1960s by studies extolling the virtues of attribute analysis. From the 1970's to the present, research has predominantly focused on artefact function and raw material availability. These approaches have rarely linked assemblage composition to the temporal characteristics of artefact discard (but see Brid 1985).

Early Studies

Interest in stone artefacts during the late 19th and early 20th centuries was mostly confined to amateur collectors who spent their leisure time scouring the landscape in search of artefacts (Griffiths 1996). These activities were largely curtailed by the implementation during the late 1960s and 1970s of State and Commonwealth legislation protecting Aboriginal archaeological sites. Pioneering Australian archaeologists Hale and Tindale undertook the first excavation of a stratified rockshelter deposit in Australia at Devon Downs in the lower Murray River valley in 1929 (Hale and Tindale 1930). Importantly, the excavation yielded a series of artefact assemblages from a stratified context, thus for the first time providing a temporal framework for the study of Australian stone artefacts.

Two prominent typological schemes emerged during the 1930s and 1940s. Tindale, based on the excavation of the Devon Downs rockshelter and the Tartangan Island open site on the Lower Murray River, proposed a four-phase model of cultural succession, defined according to the presence or absence of specific tool types (Hale and Tindale 1930:203). A fifth and older phase, the Kartan, identified from research on Kangaroo Island was added in 1931 (Tindale 1941:145). In contrast to Tindale, who was interested in establishing an Australia-wide typological sequence, McCarthy focused on defining regional sequences. It was in this context that McCarthy defined the Eastern Regional Sequence from excavations at Lapstone Creek in eastern New South Wales during the 1930's. The two phases that made up the sequence were the Bondaian and the Eloueran (McCarthy 1949:311).

Mulvaney (1961) largely discredited the typological schemes of Tindale and McCarthy. As an alternative, he adopted the method of attribute analysis within a typological framework. In his analysis of the Kenniff Cave assemblage from Queensland, Mulvaney rejected the notion of discrete types, instead opting to study continuous variability in different attribute types. This was achieved through statistical analysis of artefact size. Mulvaney (1969:106-107) proposed a scheme encompassing two distinct phases of artefact manufacture during the last 6,000 years in the Kenniff Cave assemblage, the earlier Adaptive (non-hafted) and the later Inventive (hafted). Mulvaney's innovative approach to artefact analysis was somewhat ahead of its time. Studies of function and the influence of raw material availability and distribution largely replaced Mulvaney's approach in the 1970s.

Function

In Australia, the study of artefact function offered an alternative interpretative framework for the explanation of assemblage variability to those previously used. Functional studies were linked to investigations of artefact morphology, which if successfully linked to function, could be used to identify functionally specific sites within a broader settlement system (Holdaway and Stern 2004). Early functional studies were mainly based on ethnoarchaeological observations of contemporary artefact use, usually among Aboriginal groups in the arid zone (e.g., Hayden 1977, 1979). However, Hayden (1977) was unable to correlate particular artefact morphologies with specific functions, instead observing that several different types of edge modification performed the same task. Hayden (1977) also noted that the morphology of tools changed through the process of resharpening and that one type of edge modification was often resharpened into a different form. Hayden (1977:185) observed that one particular flake tool with a notch was later resharpened into a scraper. Despite a well-established acknowledgement of the significance of resharpening on artefact morphology overseas (e.g., Dibble 1987, Jelinek 1976), a small number of Australian archaeologists are now investigating the relationship between resharpening and artefact morphology (e.g., Clarkson 2002, Hiscock 1994, Hiscock and Attenbrow 2002).

Cane (1984) attempted to identify a correlation between artefact morphology and function based upon ethnographic observations of stone tool manufacture and use in the Western Desert. He (1984) argued that artefact morphology is a reliable indication of function and that the proportion of different artefact types within an assemblage reflects the economic activities performed at a location. Like Hayden (1977), Cane (1984) observed that the majority of tools were used for woodworking and that a particular type of tool or edge modification could undertake a range of different woodworking activities. He (1984:277) noted that tula adzes were used for heavy duty woodworking and the gentle scraping and engraving of wood. Further complicating the link between morphology and function is the observation that unifacial points were used for engraving and in ceremonies and sorcery (Cane 1984:278). Although Cane (1984) notes the multifunctional characteristics of tools, he concludes that assemblage composition largely reflects site function.

By the mid 1980s, studies of artefact function turned to use-wear and residue analysis (Fullagar 1986). These studies (Kamminga 1982, 1985, McBryde 1985, McDonald *et al.* 1994) confirmed the difficulties of linking artefact morphology to a specific function. In an investigation of use-wear on horsehoof cores, burins, polished flakes and eloueras, Kamminga (1982) concluded that only the primary function of eloueras could be determined. McBryde (1985) argued that breakage patterns and edge damage indicated that backed blades from the New England area of New South Wales had multiple functions ranging from projectile points to skin piercers. McDonald *et al.* (1994) applied a similar argument to the interpretation of the function of backed blades from sites on the Cumberland Plain in New South Wales.

Fullagar (1994) states that the results of these studies have been limited by the difficulties of determining the species of the worked materials, although he points to the work of Loy (1993) as an advance in this area. Hiscock and Clarkson (2000) argue that use-wear and residue analysis is an important line of investigation, but a number of issues associated with the technique remain unresolved. They stress that tool edges frequently only retain residue from the most recent episode of use. As indicated above, tools are likely to be multifunctional and used on multiple occasions, therefore residue analysis only informs about the last task undertaken. The preservation of residue is also affected by the characteristics of the raw material and the sedimentary environment in which the artefacts were discarded (Hiscock and Clarkson 2000). One of the major benefits of use-wear studies has been the confirmation of Hayden's (1977) assertion that unretouched flakes were often used as tools, although the proportion of these in assemblages has been argued to be low (Cane 1984:276).

The results of use-wear and residue studies are certainly interesting, but they do not consistently point to an irrefutable link between morphology and function. In addition, detailed study of all artefact edges in assemblages containing thousands of pieces is not practical and is complicated by resharpening and reuse. The results of edge residue studies have rarely been extrapolated to the wider interpretation of assemblage composition. Debitage is the major component of Australian lithic assemblages, so considerable amounts of this material may not be associated with tool production. It is possible that tools were deposited independent of debitage. Therefore, it is problematic to draw links between tool function, debitage and place function.

Raw Materials and Distance to Source

The relationship between stone artefact assemblage composition and the location of raw material sources has been widely applied to the interpretation of interassemblage-variability in Australia (Byrne 1980, Cane 1984, Hayden 1977, Hiscock 1986, O'Connell 1977) and other parts of the world (Gamble 1999, Montet-White 1988, 1991, Newman 1994). Commonly known as distance decay models, these are based on the idea that differences in access to raw materials and the physical characteristics of these sources are important variables influencing artefact form (Hiscock 1986, O'Connell 1977). In Australia, these models have primarily been used to explain the economic context of the movement of stone across landscapes. Some researchers (Gould 1980, McBryde 1984, 1997) emphasise the social rather than utilitarian context of raw material movement, arguing for the existence of large-scale trade and exchange networks in the movement of raw materials and artefacts in Australia, particularly ochre, greenstone axes and grindstones.

The origins of Australian distance decay models can be found in the ethnoarchaeological investigations of raw material movement conducted in the arid zone during the 1970s (Hayden 1977, O'Connell 1977). These studies utilised knowledge of the distribution of raw material sources, artefact size and the proportion of artefact and raw material types to investigate the effects of raw material distribution on assemblage composition. They typically argue that, as the distance from raw material sources becomes greater, the degree of reduction and conservation of lithic material increases. This is reflected in a decrease in artefact size and the proportion of cortical artefacts, change in form, and different proportions of artefact and raw material types within assemblages (O'Connell 1977).

Distance decay became the driving force behind Australian stone artefact studies during the 1980s (Holdaway and Stern 2004). Byrne (1980) applied the variables identified by O'Connell to the interpretation of surface stone artefact assemblages recorded within a 27 kilometres radius of a silcrete quarry in the Murchison River region of Western Australia. Byrne's (1980) conclusions closely followed O'Connell's, noting that, as the distance between the quarry and the artefact scatters became greater, artefact size decreased and the proportion of both artefact types and raw materials changed. Other studies (Hiscock 1986, McNiven 1993) noted similar trends in assemblage composition. Hiscock (1986) argued that the location of stone sources and the characteristics of these materials determine the morphology of artefacts. He concluded from a study of assemblage composition at Lawn Hill in Queensland that raw material conservation techniques are more frequent with increasing distance from raw material sources. McNiven (1993) noted that as the distance to source increases in sites along Teewah Beach in Queensland, tool maintenance increased in importance over tool manufacture.

In recent years a number of reduction models have acknowledged that the relationship between raw material location and the intensity of reduction is not always linear (Close 1999, Cosgrove 2000, Hiscock and Clarkson 2000, Kuhn 1995). Raw materials can be moved around the landscape in many different directions. In a study of Early Neolithic assemblages from southwest Egypt, Close (1999) demonstrated that reduction in artefact size does not always correlate with increasing distance from source, especially in specialised technological systems. Kuhn (1995) argued that distance decay relationships are not manifest in all aspects of assemblage composition. From a study of Italian Mousterian assemblages, he argued that the intensity of core reduction is closely correlated with distance, however blank production technology and the intensity and frequency of retouch are independent of raw material availability. Similar variation in the intensity of reduction has been noted in Australian assemblages. Although arguing that distance decay was the over-riding factor explaining assemblage composition at Teewah Beach, McNiven (1993) noted that the linear relationship between decreased artefact size with increased distance from source did not always hold true. He suggests that this may reflect non-linear settlement and subsistence systems operating in less patterned resource areas. Hiscock (1996) observed that distance decay relationships are not always evident, especially in contexts where raw material sources are ubiquitous. Moreover, Hiscock and Clarkson (2000) note that there is no fixed rate at which assemblages change with varying distance to lithic sources. An example is the JSN assemblage from the Strzelecki dunefield (Smith *et al.* 1991). Despite the fact that the nearest source of flake quality stone is located 40 kilometres away, the presence of large artefacts, and cortical flakes and cores indicate that the JSN assemblage has none of the hallmarks of raw material rationing associated with distance decay (Smith *et al.* 1991:187).

Often, distance decay models have focused on human-environment interactions rather than the economic choices of raw material selection. This has resulted in the criticism that economic models simplify the investigation of assemblage composition and reduce the possibility of identifying change through time (Holdaway 1995a). Clearly distance to source is important (Kuhn 1995), however a range of other factors also influence assemblage composition. An alternative to distance decay is to consider the choices people made between locally available and distant raw materials and the types of artefacts manufactured from each. This can be incorporated into an investigation of the relationship between mobility and the intensity of raw material utilisation.

Distance decay models provided Australian archaeologists with an interpretation of assemblage composition independent from those that stressed style and function. However, distance decay models have often been emphasised at the expense of alternative explanations of assemblage composition and variability.

Raw Material Procurement

The methodological uniformitarian application of information derived from ethno-archaeological studies provides the context for investigating the economic context of raw material procurement and utilisation. The procurement of raw materials is a key element of hunter-gather organisation and scheduling decisions, and is important because of the high risk associated with shortfalls of material inhibiting the successful performance of subsistence activities. Two strategies, embedded and non-embedded procurement, are prominent in discussions of raw material acquisition (Seeman 1994). Embedded procurement is the acquisition of raw materials during the course of other subsistence activities, e.g., the collection of stone during a hunting excursion (Binford 1979). Binford argues that specific excursions to obtain lithic materials were rarely made and the costs associated with unsuccessful hunting or foraging excursions may be offset by the procurement of raw materials (Binford 1979). Non-embedded procurement is the acquisition of raw materials during direct and specific visits to sources. Seeman (1994) suggests that direct procurement strategies were a key feature of planning, particularly in anticipation of aggregated gatherings. In Australian ethnographic studies non-embedded strategies have been linked to large-scale trade and exchange networks (McBryde 1984, 1997). Identifying the archaeological correlates of a specific procurement strategy can be problematic.

An alternative is to investigate the relationship between mobility patterns and raw material utilisation. Gould (1978) suggested that differences existed between the use of quarried and non-quarried raw materials in the Western

Desert. He argued that quarried sources were often of a higher quality than non-quarried sources, and that more effort was invested into the procurement, transport and extension of the use-life of artefacts made on quarried stone compared to the casual procurement of non-quarried stone. In applying this model Gould (1978) discovered that tula adzes made on non-local (termed exotic) materials had no utilitarian advantage over those made on local raw materials. Therefore, to explain the presence of tula adzes made on non-local materials, Gould (1978) argued that these represent high value items with significant procurement costs. Binford (1979) disagreed with Gould's interpretation, arguing that the presence of materials from distant sources indicates the mobility distance over which subsistence activities were undertaken, and that this does not represent a significant increase in the amount of effort expended in the procurement of these materials. In reply, Gould (1980:157) suggested that the Puntutjarpa assemblage is evidence for the existence of long-distance social networks based upon the exchange and sharing of resources. McBryde (1997) offers a similar interpretation for the widespread distribution of sandstone grinding slabs from sources in Central Australia, arguing that this is indicative of social networks of trade and exchange operating over large areas. Gould and Saggers (1985) argue that the presence of tula adzes made on non-local materials in the James Range East assemblage from Central Australia, reflect technological factors associated with edge-holding efficiency influencing raw material procurement strategies. They conclude that the technological benefit gained from non-local higher quality materials may have overridden problems encountered with the transport of materials from distant sources.

In a re-evaluation of Gould's (1980) interpretation of the Puntutjarpa assemblage, Bird (1985), like Binford (1979), suggested that the presence of non-local raw materials in the assemblage might represent the procurement of materials during normal mobility strategies. Bird argued that consideration of stone procurement only in terms of tools is a weakness of Gould's model because tools account for a small percentage of artefacts entering the archaeological record. In addition, she noted that tula adzes provide a misleading account of stone procurement because they have long use-lives and are therefore likely to be discarded at a considerable distance from their source. Bird (1985) concluded that the presence of tula adzes made on non-local materials represents the relationship between artefact use-life, occupation duration and discard rather than social networks of trade and exchange. Under arid zone mobility strategies, occupation duration will rarely exceed tula adze use-life, resulting in adzes being carried between multiple locations before discard (Bird 1985).

Gamble (1993) investigated the social and spatial dimensions of stone procurement in terms of local and exotic resources, arguing that the scale at which procurement systems operated is a reflection of social life and social landscapes. Local and exotic are relative terms, the definition of which depends on the scale of mobility patterns and the composition of resources areas (Bird 1985, Dibble *et al.* 1995, Gamble 1993, Kuhn 1995:127). Gamble (1993) argues that there is no single spatial threshold between local and exotic, rather the difference between them must be specified in terms of exclusive and inclusive social contexts. In resource rich areas, social systems are exclusive, distances are constrained and the local area is small, therefore exotic items are those that have been obtained from outside the local area (Gamble 1993). In poor areas, social systems are inclusive, and the distances covered to obtain items are far greater because the local area is large, thus distance is a poor indication of exotic materials (Gamble 1993). He argues that the social context of inclusive procurement describes the exchange of materials over large distances and the increased rationing of these with increased distance from source in resource poor areas. However, he (1993:40) arrives at a similar conclusion for the exchange of raw materials in resource rich areas, stating "at the level of stone procurement there are no clear indications of exclusivity other than the restriction of the spatial scales involved".

Clearly the relationship between social systems and raw material procurement is difficult to define in specific terms. Gamble (1993) attempts to clarify this by applying his model as an explanation for the appearance of trade and exchange in prestige goods. In support, he cites McBryde's (1984) interpretation of the distribution of greenstone axes throughout southeastern Australia as representing a social network of trade and exchange. Gamble argues that the presence of greenstone axes among groups with exclusive social systems along the Murray River indicates that exotic items represented the payment of membership dues from outside groups. In inclusive social systems, exotic items represent trade as an insurance policy, but he is not explicit about the potential risks that exotic items insure against.

Gamble's conclusions are underpinned by the notion that raw material procurement is linked to exchange, but this need not be the case. As Binford (1978) and later Bird (1985) have argued, exchange is not a prerequisite to the movement of materials over large distances, these may be procured and transported during normal mobility strategies. Therefore, the presence and rationing of materials in different resource areas is better explained by other factors related to mobility and occupation duration that need not be indicative of either wider inclusive or exclusive social systems (e.g., Rolland and Dibble 1990). McBryde's (1984) model is one of many plausible explanations of the distribution of greenstone axes. Her model is based on a combination of 19th century historical accounts, linguistic data and archaeological evidence. The temporal context of McBryde's model is unknown and this has important implications for the development of exclusive and inclusive social systems, which, Gamble (1993) speculates, was most likely abrupt and punctuated. Recognising the archaeological signatures of these systems is problematic without a temporal context, especially if such systems appear late in the archaeological record. Allen (1997) for example suggested that the major increase in the distribution of Leilira blades 500 years ago in northern Australia is on the margin of archaeological visibility. Further, if trade and exchange systems are linked to resource conditions, then understanding the temporal character of the environmental context is important, particularly as this is not constant.

Gamble's model is based on a number of generalisations regarding ecological equilibrium and resultant resource stability. In Australia, ecological conditions change over many varied time frames, including both those classified as short and long-term. Indeed the availability of stone sources may change through time in accordance with geomorphological processes or overuse (Andrefsky 1994, Kuhn 1995). Mobility patterns, social organisation and exchange networks may be sensitive to these changes. Attempts to model stone procurement must acknowledge the dynamic nature of collection strategies.

Perlès (1992) notes that raw material procurement strategies are influenced by a large number of variables ranging from abundance to tradition. These can be framed as choices regarding the acquisition of different types of materials (Perlès 1992:226). Evaluating these choices in terms of direct or embedded procurement strategies is not always possible. Dibble *et al.* (1995) suggests that the presence of distant raw materials in assemblages may imply direct procurement strategies. However, as Bird (1985) demonstrated, the use-life history of artefacts is complex and those from distant source areas may have travelled between several or more locations prior to discard. In some situations (Allen 1997, McBryde 1984), it may be possible to infer trade, but for much of the archaeological record, especially flaked stone assemblages, it is difficult to assign artefacts to a specific type of procurement strategy. Attempts to model procurement strategies from raw material reduction and composition can be problematic because a range of other factors influence the intensity of reduction and discard (Roth 2000). Distinctions between procurement from quarried (formal) and non-quarried (opportunistic) sources as proposed by Gould are not crucial to understanding the process of raw material procurement because quarried sources can be exploited on an opportunistic basis (Bird 1985:31-32). In some cases the distinction may be useful for comparing differences in raw material reduction (Holdaway *et al.* 2004). Another line of investigation is to consider the nature of raw material transport in terms of nodule form, cost, mobility, occupation duration and the intensity of reduction. These variables are associated with, and reflect the overall history of occupation (Kuhn 1991:78).

Stone Artefacts and the Investigation of Mobility

In recent years, Australian archaeologists have begun to incorporate stone artefacts into considerations of mobility strategies. Often these have been linked to apparent changes in mobility strategies between the late Holocene and earlier periods (e.g., Veth 1993). These changes have been linked to models of late Holocene economic and social intensification where it is argued that increased rates of artefact discard during the late Holocene represent population increase and more sedentary settlement systems (e.g., M. A. Smith 1988, 1996). The investigation of mobility strategies and stone artefacts varies from considerations of single artefact types (Hiscock 1996) to the composition of entire assemblages (Holdaway *et al.* 2004). The approach adopted here considers the overall composition of assemblages as important to considerations of the relationship between mobility and the occupation characteristics of individual places. Again, all of these approaches owe much to the methodological application of information gathered during ethno-archaeological studies.

Stone Artefact Studies and Mobility

McNiven (1994) and Hiscock (1996) have inferred changes in mobility strategies from variation in the frequency of single artefact types. From a study of Pleistocene assemblages from Southwest Tasmania, McNiven (1994) argued that increased numbers of thumbnail scrapers, combined with increased amounts of Darwin glass, a material from a distant source area, represented a change to more mobile settlement strategies following the Last Glacial Maximum. Following a model outlined by Parry and Kelly (1987), Hiscock (1996) argued that the frequency and degree of reduction of bipolar cores provides a measure of mobility in the Kakadu region of Northern Australia. Higher frequencies of bipolar cores and the increased reduction of these has been suggested to indicate decreased mobility because bipolar reduction is a method of extending the use-life of cores when access to raw material sources is restricted by increased sedentism (Hiscock 1996:152). Using this measure, Hiscock (1996) argued that occupation of coastal wetland sites was more sedentary compared to woodland sites. A weakness of Hiscock's analysis is that he does not present other data on the reduction of raw materials, e.g., the flake to core ratio or the non-cortical to cortical artefact ratio, both of which provide a measure of the degree of raw material reduction. Therefore, it is not known whether other aspects of the assemblages support Hiscock's argument.

The appearance of the Australian small tool tradition during the mid Holocene has been argued by Hiscock to represent a technological response to increased economic stress and risk associated with greater mobility resulting in movement into previously unoccupied regions. He argued that the fine and regular shaped implements of the Australian small tool tradition e.g., backed blades, points and tula adzes were components of a hafted technology associated with highly mobile settlement systems in an environment of rapid or constant change. These specialised toolkits enhanced the chance of successfully completing a range of tasks in a risky landscape.

Hiscock (1994) argued that these new implements conferred certain advantages upon mobile settlement strategies. They provided small, easily transported artefacts that could serve as cores to produce large numbers of flakes. Additionally, the new forms of artefacts were interpreted to be multi-functional, lending themselves to extensive resharpening that extended their use-life (Hiscock 1994:277). These attributes are valuable in situations where it is difficult to replace raw materials. A problem with the model is the widespread distribution of backed blades throughout much of continental Australia. These are often found in many varied environmental contexts. In terms of the availability and predictability of resources, some of these environments would be riskier than other environmental contexts to hunter-gather groups. Interestingly, these also include areas that were already occupied. Hiscock (1994) argues that this can be explained by environmental change increasing risk to hunter gather groups in all regions. He views the reduction in backed blade production during the last 1,000 years as a reflection of diminished risk because the pace of environmental change has slowed and people are familiar with their new surroundings. This is also reflected by a period of population increase and lowered residential mobility (Hiscock 1994:283, 285, 286). However, it is problematic to apply this model to wide areas of Australia, especially when results of intensive dating programs indicate gaps in some late Holocene chronologies that suggest long-term hiatuses and abandonment of particular localities (Holdaway *et al.* 2002a).

Assemblage Composition and Mobility

Veth's seasonally determined settlement-subsistence model for the Rudall River area of the Western Desert is framed as a series of predictions regarding site location and stone artefact assemblage composition (1993:86-87). These predictions are based upon a synthesis of ethnographic information and an expected relationship between artefact manufacture, discard and time. These are:

- Ephemeral water sites will be dominated by the early stages of core reduction. Assemblages will predominantly consist of decortification, macro (large) and linear (regular shape) flakes. Few blades and exhausted cores will be present.
- Semi-permanent water sites will mainly consist of macro, linear and blade form flakes. Decortification flakes and cores will be rare.
- Permanent water sites will have the most intensive reduction of materials with all stages of reduction represented except initial decortification. Macro, linear, micro and blade form flakes will be present, as will heavily worked and exhausted cores. Tool proportions will be dominated by tula adzes and backed pieces. The greatest range of raw materials and the largest number and highest density of artefacts is expected at these sites. These also have the highest proportions of non-local stone.

The analyses presented by Veth (1993) appear to support the predictions regarding assemblage composition and water permanency. However, in a review of Veth (1993), Holdaway (1995b, 1995c) argued that the veracity of the model is impossible to assess because, in addition to problems with sampling strategies, many of the statistical analyses employed are not appropriate to the format in which the data is presented. Holdaway (1995b) additionally notes problems with the sampling strategies employed by Veth. In particular, Holdaway (1995c) is critical of Veth's analysis of assemblage composition only between water permanency classes rather than within. He (1995c) argued that this scale of analysis masks variation between assemblages from within the same water permanency zone. The occurrence and form of raw materials is one such area of variability (Holdaway 1995c:137). To illustrate this, Holdaway (1995c) summarises the proportion of raw materials presented in Veth (1993:29-36), and this is further summarised in Table 3.1.

The data indicate a large amount of variation between local raw material proportions within assemblages from the same water permanency zone. Veth (1993:30) defines local raw material as that available within a radius of five kilometres. For example, in the permanent water zone the Nyakungayawana assemblage consists of 97.83% local raw material compared to 24.90% in the Kataru and 6.44% in the Rarki assemblages. Clearly, this does not support Veth's prediction that the proportion of non-local raw materials will be greater in assemblages from permanent water zones. Holdaway (1995b) argues that this variation reflects the nature of raw material sources and the proximity of sites in relation to these. Although the prediction by Veth that raw material reduction becomes more intensive with increased occupation duration is correct and has been noted by others (e.g., Dibble and Rolland 1992, Holdaway 2000), the methods of analysis applied to the Rudall River data do not support this interpretation (see Holdaway 1995b, 1995c).

As discussed earlier in this chapter, raw materials are acquired through the complex interplay of numerous processes and often consist of multiple collection strategies operating at the same time. The nature of raw material transport within these is a reflection of the characteristics of the lithic landscape, mobility strategies and occupation duration. For example, of the 14 assemblages reported in Veth (1993), 12 are situated within five kilometres or less of a raw material source, and often several types of stone are available within this radius. Considering the rich and

Table 3.1 Water permanency class per individual site and the proportion, and distribution of local raw materials reported by Veth (1993:29-38). Variability in the proportion of local raw materials from individual locations within the same water permanency class indicate that there is no clear relationship between increased amounts of non-local materials at locations with greater water permanency.

Water Permanency	Site	Local	Local Source
Permanent	Nyakungayawana	97.83%	1 kilometres (Silcrete)
	Rarki	6.44%	At location (Quartz)
	Kataru	24.90%	Adjacent uplands (Quartzite)
Semi-Permanent	Punamalara	NA	NA
	Mukartu	75.76%	400 metres (Chalcedony)
	West McKay Rockhole	87.76%	<5 kilometres (Silcrete)
	Nungurr	No Local Sources	No local sources
	Yulpul	29.20%	Creek gravels (Quartz, Quartzite and Chert)
	Wantili	95.36%	Gibber (Chalcedony, quartz, quartzite and silcrete)
Ephemeral	Emu Lakes	96.2%	On site (Chalcedony, chert and silcrete)
	Karlamilyi	99.55%	On Site (Quartzite)
	Pulyjalyja	72.72%	River bed (Chalcedony)
	Lalapakujarra	NA	<400 metres (Chalcedony)
	Pulyjalyja Reduction/Occupation Site	97%	On site (Chalcedony)

diverse lithic terrain of the Rudall River landscape, it is not surprising that assemblages are dominated by these raw materials. Additionally, the three assemblages that contain a low proportion of local raw material may be indicative of economic decisions reflecting the physical characteristics of the raw material. In the case of the Rarki and Kataru assemblages, only one type of local raw material is available. If these materials are of a low quality, they may have been overlooked in preference to higher quality materials from more distance sources. The same argument can be applied to the Yulupi assemblage, where although three different types of raw material are available as stream cobbles within five kilometres of the site, the overall proportion of these in the assemblage is low. This may reflect the preferential selection of potentially higher quality raw materials from outside the immediate context of the Yulupi location.

An alternative interpretation of raw material proportions can be proposed through a consideration of raw material movement, not in terms of direct or indirect procurement strategies, but rather in terms of the physical form in which materials are transported. Kuhn (1992, 1995) discusses raw material transport in terms of planning and provisioning either individuals or places e.g., caches (see Hiscock [1988] for an Australian example) in anticipation of future use and needs. Provisioning strategies can be especially beneficial in lithic poor landscapes or those where the physical properties of raw materials limit reduction strategies e.g., benefits may be gained from transporting large cores into areas where local materials only occur as small cobbles (Kuhn 1995). The costs of transporting raw materials can be high, but some of these may be offset through embedded procurement strategies. Raw materials can be transported as unworked cobbles, cores and flakes. The debitage created from further reduction will vary between each of these e.g., decortification flakes will be common when unworked cobbles are transported, but rare if flakes or heavily worked cores have been transported. Therefore, a large proportion of decortification flakes in assemblages from the ephemeral water zone in Veth's study area may represent the transport of unworked stone. The differences in assemblage composition identified by Veth may reflect differences in the form in which raw materials were transported across the landscape. This can be evaluated through a consideration of reduction at different locations relative to a range of potential raw material sources. If there are no significant differences in access to raw material sources, as is the case in Veth's study, then all stages of reduction from initial decortification through to the discard of exhausted cores should be represented in the most intensively reduced assemblages.

Holdaway *et al.* (2004) offer a different interpretation of mobility based on the analysis of assemblage composition in the Stud Creek Valley of arid northwest New South Wales. The spatial and physical attributes of stone artefacts were recorded in three sample areas within an 800 metre long section of the valley. Three different sampling approaches were used. The first involved the recording of artefacts across a continuous area encompassing different land surfaces at Stud 1. This aimed to investigate the variability in the density of stone artefact distributions across different types of land surface. The second entailed the investigation of within unit artefact variability across one land surface at Stud 2. The third applied a systematic sampling strategy to the investigation of artefact distribution over a large area between Stud 1 and 2.

Analysis indicates differences in the intensity of raw material reduction between the assemblages. Despite both quarry (microcrystalline) and non-quarry (clast) sources of silcrete being abundant within two kilometres of the assemblages, the proportion of each of these materials varies between assemblages. Microcrystalline silcrete generally occurs as outcrops on surrounding hill-slopes, while clast silcrete commonly occurs as gibber cobbles (a stone pavement) (Doelman *et al.* 2001). The proportion of clast silcrete is highest in the Stud Systematic assemblage and lowest in Stud 2. Variation in raw material proportions is also reflected in the reduction of the two types of silcrete, with the non-cortical to cortical core ratio and the non-cortical to cortical flake ratio indicating that the reduction of microcrystalline silcrete was more intensive in all three assemblages (Holdaway *et al.* 2004).

A series of other measures indicate differences in the intensity of reduction between the assemblages. Both the flake to core and the flake to tool ratios are highest for the Stud 2 assemblage and lowest for Stud Systematic in both types of silcrete, indicating that the intensity of raw material reduction was greatest at Stud 2. In addition, the average length of complete flakes is smallest in the Stud 2 assemblage and largest in the Stud Systematic assemblages providing further evidence raw materials were more intensively reduced in the Stud 2 assemblage. In-situ reduction of cobbles is indicated by a consistent reduction of flake size with decreasing proportions of cortex (Holdaway *et al.* 2004).

The Stud 2 assemblage is clearly more intensively reduced than the Stud 1 and Stud Systematic assemblages. This is interpreted to represent differences in the way that locations in the Stud Creek Valley were occupied. These do not reflect functional differences between the assemblages, but rather represent varying degrees of occupation intensity (Holdaway *et al.* 2004). The episodic nature of occupation in the Stud Creek Valley is demonstrated by the radiocarbon determinations from hearths. These indicate numerous gaps between periods of hearth construction, the longest of which is at least 200 calibrated years in duration (Holdaway *et al.* 2002a). Holdaway *et al.* (2004) argue that the differences in assemblage composition over such a small spatial area indicate the existence of assemblage types beyond those commonly defined in Australian arid zone mobility models (e.g., Veth 1993).

Economic Approaches to the Analysis of Stone Artefact Assemblage Composition

As an alternative, an economic approach to the interpretation of assemblage composition is developed in the following section. Economic approaches focus on the technological organisation of raw material utilisation and the implications of this for investigating artefact discard. Many of these ideas are derived from behavioural ecological models that emphasise the role of risk in conditioning assemblage composition. Risk can be investigated from two different perspectives in stone artefact studies: raw material abundance and tool design. In the context of western New South Wales it is concluded that the concept of risk is more appropriate to an investigation of choices relating to raw material selection and utilisation rather than exclusively to tool design alone. The methodological application of these measures provides a framework for investigating assemblage composition without recourse to substantive interpretative models.

Local versus Non-Local Raw Material

The distinction between local and imported raw materials is difficult to define. In regions, where raw materials are ubiquitous, establishing the source of particular types of stone can be all but impossible. Source characterisation (Sheppard 1997) can be useful in situations where limited sources are available. In rich lithic landscapes these studies have a lower utility because the number of potential sources of quality materials is often greater and these are widely distributed across the landscape, rather than confined to specific areas. Attempts to define local and non-local materials on the basis of standard distance measures are also problematic. For example, Veth (1993) defines local as less than five kilometres in the arid Rudall River region, with no justification provided for choosing this arbitrary figure. If this figure is defined according to mobility strategies, then it can be expected to vary between different resource zones (Gamble 1993). A more practical distinction between local and non-local raw materials is that all materials not occurring within the immediate vicinity of a site are imported (Dibble *et al.* 1995:261). Adopting this concept removes the need for an arbitrary distinction between local and non-local materials, instead focusing on the form in which materials were transported across landscapes.

The Economics of Raw Material Transport and Reduction

The abundance of different types of stone has implications for the transport of raw materials (Kuhn 1992:192). Kuhn suggests that in landscapes where raw materials are abundant, the presence of artefacts made of materials from distant sources reflects the loss or replacement of regularly transported personal gear. In landscapes where raw material sources are rare or unpredictable, transport and provisioning are more likely to be prevalent (Kuhn 1992:192).

Hunter-gather enthno-archaeological research indicates that the acquisition and transport of raw materials involves a range of costs associated with time investment and energy expenditure. These have frequently been framed within the context of different types of procurement strategies (e.g.,

Binford 1978, Gould and Saggers 1985), but as discussed, the difficulty of identifying the material correlates of specific procurement strategies often precludes a consideration of cost. Alternatively, the costs of raw material acquisition can be evaluated within a framework of risk. Elston (1990) defines two types of risk, venture and contingency. Venture risk is the probability that the costs of obtaining raw materials will exceed the benefit gained in the increased efficiency of performing subsistence activities through the use of stone artefacts (Elston 1990:154). The main costs of venture risk are those associated with the procurement and processing of raw materials and these vary according to the characteristics of the lithic landscape. Contingency risk is the probability of having insufficient tools and/or raw materials to undertake subsistence related activities (Elston 1990:154). Contingency risk increases as the consumption of raw materials and tools through reduction and resharpening increases. Short-term shortages of raw materials may result from limited access to lithic sources resulting from decreased mobility.

The transport of raw materials is an important element of cost-benefit models, especially as they relate to the physical characteristics of raw materials and their distribution across the landscape. Elston (1990) argues that in most lithic terrains, high quality materials have restricted distributions compared to those raw materials of a lower quality which are often more widely available. In this situation, the costs of acquiring high quality raw materials increase, especially if opportunities to acquire these are restricted through decreased mobility. These costs however can be offset through the utilisation of abundant lower quality materials. Therefore, assemblages will contain extensively reduced high quality raw materials and large proportions of lower quality material that are available closer to the site (Elston 1990:160).

The concept of risk has had a limited application to studies of Australian stone artefacts (Hiscock 1994, Holdaway 2000, McNiven 1994). Both Hiscock's (1994) and McNiven's (1994) models were discussed earlier in terms of mobility strategies. Of the three applications of risk, only Holdaway's (2000) analysis of assemblage composition of the Pleistocene assemblages from Bone Cave, Tasmania applied a cost-benefit analysis of raw material acquisition and reduction to the investigation of mobility and economic choice. Holdaway (2000) demonstrated that raw material reduction was more intensive in the 15,000 BP occupation than the combined 24,000 and 29,000 BP occupations. He interpreted this to indicate that the use of Bone Cave varied through time, to increased duration occupations at 15,000 BP. Following the Elston model, the more intensive utilisation of quartzite river cobbles available within the immediate context of the site during the recent occupation was argued to represent a response to the increased risk of experiencing raw material shortfalls under conditions of decreased mobility. That Holdaway (2000) does not make a distinction between venture and contingency risk suggests that the processes to which these refer may not always be mutually exclusive or readily identifiable in the archaeological record.

The spatial scale at which differences in the intensity of raw material reduction can be detected will vary between lithic landscapes. In landscapes that are poor in lithic resources of any kind, reduction of all materials is likely to be intensive because raw material conservation will be of the utmost importance, although variations form this scenario have been noted (e.g., Smith *et al.* 1991). Assemblages from areas where raw materials are widely available may not exhibit evidence of the same degree of conservation as those from source poor areas. In some cases differences in the degree of reduction between materials of varying quality have been identified in rich lithic landscapes, where the distances over which raw materials have been transported may be short (e.g., Holdaway *et al.* 2004).

Torrence (1989) argues that technology (in this case tools associated with resource procurement, specifically food) is a strategy of risk reduction. Hiscock (1994) later applied this idea in an expanded format as an explanation of the appearance of the Australian small tool tradition. Torrence (1989) defines risk as the loss of resources, which increases as the spatial and temporal variability of prey resources becomes greater, particularly in areas of high latitude where prey species are mobile. Within this framework, technology can be investigated from two perspectives. First is organisation, involving the design of time and energy budgets for the procurement, manufacture, use and repair of tools (Torrence 1989:60) e.g., raw material choice is linked to risk, in that people choose the least-cost material that suits their needs. At the same time, Torrence does not regard lithic technology as a direct reflection of the lithic landscape. Instead, she notes that a complex range of variables influences raw material procurement, a point also made by critics of distance decay models. Second is assemblage structure, which is comprised of the composition of tool types, diversity of tool types and the complexity of individual tools (Torrence 1989:60). She argues that in areas where prey species are highly mobile and the risk of failure the highest, investment in technology is greatest and specialised tools predominate (Torrence 1989:61). Nelson (1991) suggests that the relationship between technology and the representation of tools in assemblages is not the result of single strategies, but rather there are a range of technological and social choices available to people.

A problem acknowledged by Torrence (1989) is that her model is largely dependent on making functional inferences about the use of different tool types, primarily regarding food procurement strategies. Alternatively, she suggests that calculating diversity rather than reconstructing assemblage composition based on functional types may provide a method for investigating assemblage structure. Ultimately, she argues that the complexity of individual tools in terms of 'technounits' is the most appropriate measure. Following Oswalt (1976), technounits are defined as "an integrated physically distinct, and unique structural configuration that contributes to the form of a finished artefact" (Torrence 1989:61). The total or average number of technounits per assemblage increases with greater risk because multi-component tools are viewed as a risk reduction strategy (Torrence 1989:61). On one level this contradicts Torrence's previous point that tools in high-risk areas will be specialised because multi-component tools may represent non-task specific use. Considering the non-functionally specific nature of different forms of retouch noted by Hayden (1977), there is no reason to assume that multi-component artefacts are the stone equivalent of a Swiss Army Knife. Increase in the average number of retouched edges may reflect an increase in the intensity of raw material utilisation associated with an increase in occupation intensity (see Dibble 1988, Dibble and Rolland 1992). Kuhn (1994) also notes that the degree of, and type of, retouch on a tool may reflect artefact use-life history, rather than considerations of tool design, especially since by definition heavily worked tools are likely to have been transported between, and used in, several different contexts.

As Hayden *et al.* (1996) suggest, investigations of tool design within the context of risk are interesting, but all too often consideration has not been given to the wider range of components that make up lithic assemblages. This is particularly pertinent considering that tools comprise a small proportion of items within most assemblages (typically less than 10% in western New South Wales). Exclusively focusing on tools ignores the wide range of other constraints and considerations influencing raw material reduction (both core and tool) and assemblage composition. Applying the concept of risk to a wider economic perspective of raw material reduction and mobility provides a broader consideration of these factors.

Interpreting Stone Artefact Assemblage Composition in a Time-Averaged Archaeological Record

The concept of artefact discard and assemblage accumulation as a time-dependent process was introduced in Chapter One. The notion is now discussed in detail with specific reference to the implications of this for the composition of Australian arid zone stone artefact assemblages.

Stone Artefact Discard, Raw Material Reduction and Assemblage Composition

Arid zone artefact assemblages consist of both short and long use-life artefacts. Short use-life artefacts are those that have performed little or no work (Shott 1989, 1995). They are described as having a low amount of their potential utility realised (Shott 1996b). Characteristically, these are unretouched flakes and other debitage produced during core reduction and tool manufacture (Holdaway *et al.* 2004, Shiner *et. al* 2005, 2007). These have a high probability of discard and the majority of assemblages are composed of large numbers of these items. Long use-life artefacts mainly consist of retouched tools that can be ranked according to the degree to which they have been resharpened. The least resharpened tools are utilised flakes, which exhibit non-invasive retouch. These are usually flakes that have been used to perform work without prior edge modification. More heavily resharpened tools consist of denticulates, notches, scrapers and burren and tula adzes that have been deliberately modified before use. The most heavily resharpened item in arid zone assemblages is the tula adze, which following repeated resharpening are often, but not always, discarded as worked-out 'slugs' (Gould 1980:127, Hayden 1979:12, Holdaway *et al.* 2004). In some cases, the use-life of these items may exceed the occupation duration of a single location.

The discard of artefacts is a time-dependent process and the composition of assemblages is an in-direct result of occupation duration (Bamforth and Becker 2000:283). As occupation becomes longer in duration the probability that long use-life artefacts will be discarded increases (Holdaway *et al.* 2000). Therefore, tula adze slugs and heavily resharpened scrapers will be found at those locations that were the repeated focus of occupation. Contrary to Cane (1984), the relative proportion of tool forms is not a direct reflection of either artefact or site function, but is rather an indication of the duration and frequency of occupation events. Nor is assemblage composition a direct reflection of population. Estimates of arid zone populations, although attempted (Allen 1972), are problematic because there is no simple relationship between the quantity of archaeological debris and population size (Hiscock 1986:47). Complicating this, particularly over the period in which ethno-historical observations have been undertaken, is the fluid nature of arid zone group membership (e.g., Myers 1986:101). Differences in the intensity of raw material reduction offer one promising approach to the investigation of assemblage composition because increased occupation duration may lead to the more intensive utilisation of material available within the immediate context of an occupied location (Dibble 1988, Dibble and Rolland 1992, Elston 1990).

Raw material utilisation can be evaluated through the study of assemblage composition. Assemblages that reflect more intensive reduction of raw materials will exhibit high flake to core ratios, a low proportion of cortical artefacts, a decrease in debitage size and heavily worked tools and cores (Dibble *et al.* 1995:267). These characteristics, combined with the concept of time-dependent artefact discard, provide a method for investigating the intensity of landscape use from stone artefact assemblages without recourse to functional interpretations. Locations that were the focus of repeated occupation will contain large numbers of long use-life artefacts and intensively utilised raw materials.

Occupation duration is not an absolute, it can only be quantified in a relative sense, and therefore intensity of occupation is a more appropriate term for the description of artefact discard. Assemblages that contain a high proportion of long use-life artefacts indicate intensively utilised locations compared to other places where ephemeral occupation is indicated by the presence of few long use-life artefacts. To investigate this requires that the life histories of artefacts be evaluated from procurement to discard.

Summary

In this chapter, it has been argued that Australian archaeologists have rarely applied analytical techniques, which permit the effective investigation of stone artefact assemblage composition within a framework of landuse and mobility. Those who have applied distance to source models will disagree with this statement, but, as discussed above, such models have all too often been framed as linear patterns of movement in raw materials and people. Nor have functional studies fared any better; since the multifunctional nature of Australian flaked stone artefacts precludes the identification of specialised activity areas at a landscape scale. This, combined with the problems of defining typological site types discussed in Chapter Two, suggests that function be discarded as the dominant interpretative paradigm in Australian hunter-gather archaeology.

An alternative approach based on the analysis of assemblage composition from an economic perspective is proposed. The main components of this are the investigation of raw material availability and transport, as well as utilisation and discard within a time dependent model. This model predicts that different conditions of raw material availability, combined with varying degrees of occupation intensity, will lead to particular characteristics within the composition of lithic assemblages. The time-dependent nature of assemblage formation allows the investigation of long-term place use histories and the factors that structure the reduction and discard of stone artefacts. These provide an independent method for evaluating the temporal character of the factors that structure assemblage formation. The key is to develop a temporal framework within which a comparison of the variability between assemblages can be undertaken.

These principles are applied in the analysis of the Pine Point-Langwell assemblages in later chapters. This provides a means of incorporating detailed assemblage analyses into considerations of long-term accumulated place use histories. The following chapter builds towards this goal through the discussion of the archaeological context of the study in western New South Wales and the dynamic nature of landscape change in this region. More specifically, the implications of this for developing behavioural interpretations of material from deflated contexts are highlighted.

Chapter Four: Environmental Context of the Study Area

Introduction

The present environment is frequently used as an analogue for the environment during the past. In some instances, features of the modern environment e.g., water availability and rainfall are used by archaeologists to predict conditions during the past and are incorporated into interpretations of the spatial distribution and composition of the archaeological record. Quantifying the range of flora and fauna resources available to past inhabitants and the role of these in logistical strategies is a key element of ecological approaches (e.g., Barton 2001). The fact that the current environment may not mirror that of the past is often overlooked in these approaches. This is particularly so in landscapes with a recent history of pastoral land use where significant changes to flora and fauna distribution may have occurred.

The environmental context of the present study is not presented as a pretext for establishing settlement/subsistence models. Environmental change, particularly that has accompanied the shift from hunter-gather to pastoral landuse during the last 150 years (Fanning 1999, Pickard 1994), indicates that the present environment of western New South Wales is not an appropriate analogue for the past. Instead, much of the following chapter builds on the model of archaeological deposit formation discussed in Chapter One by focusing on the geomorphological history of arid western New South Wales and the extant context of the study localities across the Pine Point-Langwell study area. Particular attention is given to the role of geomorphic processes in the formation and exposure of the surface archaeological record at these locations. The implications of these for investigating and comparing the occupational history of the locations are discussed in detail, especially in terms of the importance of establishing chronological contexts for deflated deposits.

The Environment of Far Western New South Wales

Far western New South Wales is situated on the southeast margin of the Australian arid zone. Mabbutt (1971:66) defines the Australian arid zone as areas with "an excess of potential evaporation over precipitation". Based on this criterion, approximately 70 per cent of the Australian continent is presently classified as arid (Mabbutt 1971:66). The Australian arid zone extends from northern Victoria in the south to the northwest coast of Western Australia and incorporates a diverse range of landforms and environments. The current average annual rainfall of 250 mm places the Pine Point-Langwell study area on the present boundary between the arid and semi-arid zones.

Far western New South Wales extends from Cobar in the east to the South Australian border in the west and from the Murray River in the south to the Queensland border in the north. This extensive region contains a number of different environments. These range from riverine flats and clay plains along the Murray and Darling Rivers characterised by river red gum (*Eucalyptus camaldulensis*) woodland along the river and black box (*Eucalyptus largiflorens*) woodland on the margins of the floodplain, and mallee country on the red linear dunes. Upland and pedimont range systems occur within the foothills and high plains of the Barrier Range dominated by presently sparse mulga (*Acacia aneura*) woodland and numerous ephemeral creeks with galleries of river red gum (*Eucalyptus camaldulensis*). Stony desert dominated by gibber pavements of fist sized stone nodules and mesas with low relief occur in the far north. Ephemeral lakes are common along the Murray-Darling River System and isolated lakes and box swamps within small internal drainage systems occur across much of the region. Aridity, as measured by annual rainfall and rates of evaporation, generally increases from the southeast to the northwest. The following section describes previous archaeological research in western New South Wales and outlines the relationship between recent geomorphic processes and the surface archaeological record.

Archaeological Context of Western New South Wales

A large amount of archaeological research, relative to other parts of Australia has been undertaken in western New South Wales. The results of this research reflect the complex and diverse archaeological record of the region extending into the late Pleistocene. A brief historical review of this research provides a general archaeological background for the present study.

Willandra Lakes

The Willandra Lakes are etched into the history of Australian archaeological research because of the archaeological and geomorphological investigations undertaken during the late 1960s and early 1970s. Excavation of a carbon encrusted and mineralised scatter of burnt human bone and stone artefacts on the Lake Mungo Lunette (Mungo 1) established a late Pleistocene date for the occupation of western New South Wales (Bowler *et al.* 1970:46). Based upon stratigraphy, the burnt human bone was estimated to date to between 25,000 and 32,000 years BP (Bowler *et al.* 1970:46). Later results confirmed that the Mungo 1 burial dates within the range of 25,500 ± 1,000 years BP (Bowler *et al.* 1972:50). The stone artefacts from Mungo 1 were the initial assemblage from which the Australian core tool and scraper tradition was defined (Bowler *et al.* 1970). Later excavations at Mungo 1 by John Mulvaney confirmed the integrity of association between the surface stone artefacts and the Mungo stratigraphic unit within which the cremation was intrusive (Shawcross 1998:187). Further human remains were discovered in the Willandra Lakes and these became important in wider debates about the origins of Aboriginal populations (Bowler and Thorne 1976, P. Brown 1987, Pardoe 1991, Thorne 1976, 1977, S. Webb 1989). New debates about the age of some of the skeletal remains from Lake Mungo have emerged in the last few years (Bowler and Magee 2000, Thorne *et al.* 1999). Recent studies of skeletal remains across western New South Wales have investigated regional variation in burial practice (Littleton 1998, 2002 Pardoe 1993a), burial taphonomy (Littleton 2000) and the development of cemeteries (Littleton 2002, Pardoe 1988, 1993b).

Allen (1972, 1974, 1990, 1998) undertook a wider investigation of the distribution of archaeological material throughout the Willandra Lakes and neighbouring areas as part of his doctoral research in the late 1960s. The primary aim of Allen's research was to "try and fit the Willandra material into a wider regional sequence, rather than to do a detailed settlement study that would be in an archaeological limbo as far as the rest of Australia was concerned" (Allen 1972:135). To achieve this, Allen (1972) compared data from the Willandra Lakes to other sites from Lake Tandou and Burkes Cave to the northwest.

Allen (1972:356) concluded that the contents of the faunal and stone artefact assemblages combined with ethnographic data suggest "...a single continuous cultural tradition...". Recently, Allen (1998) challenged the notion of continuity developed in his thesis, suggesting that the technological framework he adopted was unable to properly address Aboriginal subsistence in terms of continuity or change. More recent work in the Willandra Lakes has investigated the role of lacustrine resources in Pleistocene diets (Johnston 1993). As discussed in the previous chapter, Hiscock and Allen (2000) re-evaulated the Lake Mungo stone artefact assemblages, arguing for a greater degree of variability than previously identified. In particular, they analysed the assemblages in relation to the location of quarried silcrete outcrops on the Lake Mungo backshore.

Menindee Lakes

Investigations during the early 1980s of middens at Lake Tandou on the Lower Darling River provided further evidence of Pleistocene occupation (Hope *et al.* 1983). Radiocarbon determinations from eleven middens fall into two phases; one group of seven sites between 27,000 and 22,000 years BP with a later group of four dated to between 16,000 and 15,000 years BP (Hope *et al.* 1983:48-49). Hope (1993) argued that the two phases identified at Lake Tandou are supported by Pleistocene dates from middens in other areas of southwest New South Wales. The absence of middens between 22,000 and 17,000 years BP in the Willandra Lakes may reflect the drying of the Lakes (Hope 1993:191). However, this explanation is not applicable to Lake Tandou, where uniform sedimentation between 27,000 and 14,000 years BP indicates stable lake-full conditions (Hope 1993:191). Hope (1993) was unable to offer an explanation for the 5,000 year gap in the midden chronology at Lake Tandou.

Balme (1995) examined the composition of shell middens spanning 30,000 years from the Lower Darling River with the aim of producing a regional synthesis of subsistence patterns. Following Allen (1974), she (1995:18) concluded that "...apart from some mammalian extinctions, the range of aquatic foods used and the technologies for gathering them, appear to have remained essentially unchanged over the last 27,000 years in the Lower Darling region...". Despite continuity in the contents of middens over a 30,000 year period, Balme notes that there is an increase in the number and complexity of middens during the Holocene (Balme 1995:18). She argues that the larger and more complex Holocene middens reflect increased occupation duration by perhaps larger groups of people. Ultimately, she (1995:18) viewed the development of these sites as the establishment of "...social control mechanisms for the exploitation of the highly productive, but geographically constrained riverine resources...".

Stone Artefacts

Despite the large amount of archaeological research undertaken in western New South Wales detailed studies of stone artefact assemblages are rare. A number of papers summarise the analysis of the stone artefact assemblages studied by the Western New South Wales Archaeology Programme (Holdaway *et al.* 2000, Holdaway *et al.* 2004, Shiner *et al.* 2005, 2007) the results of which are discussed throughout this study. Cathy Webb's doctoral research in the Menindee Lakes during the late 1980s also provides detailed studies of stone artefact assemblages. Webb (1993, 1996) investigated the role of raw material transport in the formation of surface stone artefact assemblages from Kinchega National Park at the Menindee Lakes. The sand plain of Kinchega contains no natural stone sources. The nearest sources of stone artefact quality raw material occurr in the Scopes Range some 40 kilometres to the northwest of the Menindee Lakes area (C. Webb 1993:105). During surveys at Kinchega, Webb (1993:105) noted that many large isolated cobbles of silcrete, sandstone, quartzite and quartz occur throughout the National Park. She argued that the importation of cobbles commenced during the late Pleistocene and persisted until the mid-Holocene and that the cobbles acted as portable quarries positioned across the landscape where there was the possible future need for stone.

Witter (1992) investigated stone artefact assemblages in the Tibooburra region in the north, comparing these to assemblages from further to the east along a transect stretching from Tibooburra in the northwest through Cobar in central western New South Wales to Boorowa in the east. Witter (1992) demonstrated that assemblages in the north were primarily flake tool based, with tula, pirri and microblade forms common. Raw materials sources are abundant with silcrete derived from outcrops and gibber pavements dominating the assemblages. Quartz was also present in lower proportions, especially in the ranges as outcrops (Witter 1992:134).

Late Holocene Archaeology of Far Western New South Wales

Excluding the research conducted as part of the Western New South Wales Archaeological Programme, the majority of investigations into the Holocene archaeology of western New South Wales have focused on the existing Darling River channel and adjacent lakes. Despite this research, Balme and Hope (1990) note that few locations with Holocene archaeological materials have been recorded along the present course of the Darling River. They suggest that both research bias and poor exposure account for this situation (Balme and Hope 1990, Balme 1995:19). In other areas large numbers of surface stone artefact scatters of inferred mid to late Holocene age (based on stone tool typology) have been recorded during cultural heritage management surveys (Martin 1985a, 1985b, Rich 1986, Witter 1989). However, the details of these studies remain unpublished.

Previous Archaeological Fieldwork in the Study Area

There has been very little archaeological research conducted within the boundaries of the study area prior to the present project. An archaeological survey for the Mildura to Broken Hill powerline during 1977 covered a section of the study area encompassing Pine and Rantyga Creeks (McIntyre 1977, 1981). McIntyre (1981) defined three different survey units relevant to the study area:

- South Ita Road/Pine Creek – The northern portion of this unit covers the treeless plain south of Pine Creek and includes numerous internal drainage basins expressed as small swamps.
- Pine Creek – This unit intersects the junction of Pine and Rantyga Creeks.
- Pine Creek/Silver City Highway – This includes the gently sloping to undulating terrain through which Rantyga Creek flows.

McIntyre (1981:28) recorded a total of seven archaeological sites along the survey section of Pine and Rantyga Creeks and eight sites on the internal drainage plain south of Pine Creek. There is no specific information on the composition of these sites beyond the observation that the assemblages consist primarily of scattered distributions of silcrete and quartz flakes with occasional grinding gear. However, it was noted that one of the most extensive sites recorded during the survey was located near the Pine and Rantyga Creek confluence. In addition, McIntyre regarded the discovery of large sites on the internal drainage plain south of Pine Creek as a surprise because of the absence of large creeks. No worked outcrops of stone were recorded during the survey, but he did note that rounded quartz nodules were widely available across the plain. The results of this survey provide useful background information on the likely distribution of land surfaces and the landform context upon which surface archaeological distributions will be visible in the study area.

Geomorphology and Landscape History of Western New South Wales

The formation of the surface archaeological record reflects the operation of cultural and environmental processes over a range of temporal and spatial scales. The role of geomorphic processes in the formation of deflated surface archaeological deposits was discussed with specific reference to western New South Wales in Chapter One. A consideration of the broader landscape backdrop upon which the archaeological record has formed heightens an understanding of the relationship between these processes. The following section describes the major landform units that occur across the western New South Wales region, and in the vicinity of the Pine Point-Langwell study area.

The Ranges

The study area is situated near the boundary of the North Western Fold Belt and the Murray Basin geological provinces. The North Western Fold Belt consists of the three major rangeland regions of western New South Wales; the Barrier Range extending from Broken Hill in the south to Fowlers Gap in the north, the Scope-Bynguano Range to the east of Broken Hill and the Grey Range in the far north. The southern section of the Barrier Range, which is the closest of the ranges to the study area, is located 30 kilometres north of the Pine Point-Langwell Study area. Mount Robe at 474 metres above sea level and located approximately 50 kilometres to the north of Broken Hill is the highest elevation within the Barrier Range.

Most of the southern region of the Barrier Range comprises Precambrian granite gneisses, schist, pegmatite and limestone of the Broken Hill Block (Wasson 1979:27, Cunningham *et al.* 1992:13). The Broken Hill Block is famous for its rich deposits of zinc, silver and lead. Small outcrops of Devonian sandstone, siltstone and conglomerates along with isolated areas of Tertiary silcrete are also found (Cunningham *et al.* 1992:13, Burton 1994:5). The geological history of the Broken Hill Block indicates major periods of erosion during the past (Table 4.1).

Plains and Alluvial Fans

To the southeast of Broken Hill, the Redan Fault separates the North Western Fold Belt and the Murray Basin (Stevens 1986:75). Quaternary sediments cover much of the Murray Basin with level to gently undulating plains

Table 4.1 Description of the geological history of the Broken Hill Block (Adapted from Stevens 1986:75-78).

Date	Geological Event
1800 MYA	Deposition of the sedimentary/volcanic sequence of the Willyama Complex of the Broken Hill Block
1490 - 1100 MYA	Uplift of rocks presently seen at the surface
1100 - 500 MYA	Erosion, weathering, faulting, metamorphism and igneous intrusion of the uplifted sediments
500 MYA - Present	Weathering, erosion, minor fault movements and possibly the development of vein-type mineral deposits and the formation of Tertiary silcrete

and dunes of red and brown clayey sand, loam and lateritic soils predominating in the northern and western section. Quaternary alluvial fans, floodplains, outwash and drainage flat deposits of black and red clayey silt and sand are found adjacent to the major watercourses in this area, e.g., Pine Creek and Stephens Creek (Cunningham *et al.* 1992:14). The Riverine Plain, which dominates the southern and eastern section of the basin, is comprised of Quaternary alluvial deposits with playas, lakes and claypans of black and grey clay and silt deposits (Bowler 1986).

The alluvial fans in the northern and western section of the Murray basin evidence the deposition of sediments from the Broken Hill Block sometime between 23 to 5 million years ago (MYA) and 30,000 to 16,000 years BP (Stevens 1986:95). A general model for the evolution of the Quaternary alluvial fans of the Broken Hill district was developed by Wasson (1979), who described the evolution of alluvial fans on the Mundi Mundi Plain to the west of Broken Hill. He proposed four depositional units: Umberumberka, Korkora, Mundi Mundi and Thackaringa. Deposition of the Umberumberka unit was caused by a major erosional event in the Broken Hill Block attributed to a wetter climate during a period of depressed temperatures between 30,000 and 16,000 years BP (Wasson 1979:21). The Korkora unit was deposited on top of the Umberumberka unit by shallow channels (Wasson 1979:21). Increased rainfall between 6,000 and 3,000 years BP resulted in the deposition of the Mundi Mundi unit followed by stream dissection of the alluvial fans caused by reduced rainfall at 3,000 years BP (Wasson 1979:21). A slight increase in rainfall at 1,000 years BP resulted in the accumulation of sediments (the Thackaringa unit) in the trenches of the dissecting streams (Wasson 1979:21). Wasson (1979) concluded that climate change was one of the major factors influencing the development of the Mundi Mundi Plain.

Drainage

Warner (1986:50) describes three types of Australian arid zone surface drainage organisation:

1. Co-ordinated internal drainage with external outlet – this has a patterned and predictable distribution of water resources with an outlet flowing into the ocean.
2. Co-ordinated internal drainage with no external outlet – also has a patterned and predictable distribution of water resources.
3. Un-coordinated drainage – disconnected drainage with little pattern in the distribution of water resources across the landscape.

The Barrier Range forms the watershed between the Murray-Darling, Lake Eyre and Bulloo-Bancannia drainage divisions (Warner 1986:50). All watercourses draining to the east of the southern region of the Barrier Range belong to the Murray-Darling drainage division, a co-ordinated internal drainage basin with an external outlet on the South Australian coast (Warner 1986:50). Those draining to the east in the northern Barrier Range belong to the Bulloo-Bancannia drainage division, an uncoordinated internal drainage system (Dunkerley 1992:417). Watercourses draining the western flank of the range flow towards Lake Frome in the Lake Eyre drainage division, a co-ordinated internal drainage system without an external outlet (Dunkerley 1992:417).

The Darling River is the main watercourse in western New South Wales. The Darling receives the majority of its flow from rainfall on the Queensland highlands and therefore may contain a large amount of water even during periods of prolonged drought in western New South Wales, but may also cease to flow, instead forming a series of stagnant waterholes (Cramsie and Hawke 1984). Rapid soil absorption of rainfall means that local rainfall contributes little to the overall flow of the Darling River (Gill 1973:10). The present channel of the Darling results from an aevulsion event dating to the terminal Pleistocene (Bowler *et al.* 1978, Bowler 1986, Rognon and Williams 1977). Prior to this, an ancestral channel operated from about 36,000 years BP (Balme and Hope 1990). The ancestral channel now forms a series of anabranches (branches of a river that reconnect with the main channel), on either side of the present channel, known as the Great Darling Anabranch and Talyawalka Creek.

The many creeks that drain the Barrier Range are ephemeral and flow only after extended periods of heavy rain. The distance that water flows along these streams varies according to the intensity of rainfall within their catchment area (M. Williams *et al.* 1993:143). It is not uncommon for creeks to cease flowing before reaching a flood-out area or confluence with another watercourse (Dunkerley 1999:1577). Ephemeral creeks in the Barrier Range rarely flow. For example, creeks in the northern Barrier Range flow on average only 10 times per year (Dunkerley 1992:419).

Recent Geomorphic History of Western New South Wales

The present landscape of western New South Wales reflects the past 150 years of European pastoral land use, with the effects of sheep grazing complementing irregular rainfall and wind as the major agent of geomorphic change across the region (Fanning 1999:204, Pickard 1994:281). Sheep grazing, plus the introduction of feral herbivores (rabbits and goats) has exposed the highly dispersible soils to erosion through a reduction in available plant material and trampling (Pickard 1994:274). Further exacerbating soil erosion was the widespread felling of mulga (*Acacia aneura*) for fence posts and mining activities (Fanning 1994:176, Pickard 1994:273). Once exposed, the soils were far more susceptible to erosion by wind and water (Pickard 1994:274).

The effects of recent geomorphic change are most clearly visible within ephemeral creek systems. In upland catchments a combination of sheetwash, rilling, gullying and aeolian deflation have removed large amounts of sediment, resulting in the formation of extensive lagged areas adjacent to ephemeral creeks (Fanning 1994). Fluvial transport has deposited much of the sediment onto floodplains and floodout areas on the plains.

The deposition of large amounts of sediment in and along creeks, together with the colonisation of creek beds by river red gum trees (*Eucalyptus camaldulensis*), has altered flood discharge patterns (Graeme and Dunkerley 1993:141). Trees growing in creek beds act as barriers to water flow through the accumulation of flood debris around the trunk of the tree (Dunkerley 1992:419, Graeme and Dunkerley 1993:144). This causes flow divergence, turbulence and bed scour upstream of the obstacle. Immediately downstream of the barrier is a zone of lowered flow velocity where sediments are deposited, forming channel-islands and sand bars (Graeme and Dunkerley 1993:144). Another effect is the redirection of water flow against either one or both banks, causing scouring and a widening of the creek. Over time, this results in increased channel width relative to depth (Graeme and Dunkerley 1993:144). There is also a reduction in water carrying capacity as channels become shallower through sediment accumulation. During periods of high discharge, creeks are more likely to overtop their banks, inundating extensive areas of land and depositing large amounts of sediment.

Increased erosion since the introduction of sheep grazing has also contributed to the formation extensive scalding across much of western New South Wales. Scalds are formed by the removal of topsoil by wind and/or water erosion, exposing compact clay subsoil surfaces with low infiltration rates. Artefact distributions are highly visible on these surfaces. As the topsoil is eroded, artefacts previously encased within it conflate onto the scalded surface, forming lag on the exposed subsoil surface.

These processes are not unique to the period following European settlement, but the rate at which landscape change is occurring has accelerated during the last 150 years (Pickard 1994). This is most clearly visible in stream sedimentation rates. For example, investigation of sedimentation for Umberumberka Creek west of the Barrier Range indicates that the highest rate of sedimentation since 6000 years BP has occurred during the last 150 years (Wasson and Galloway 1986). Similar patterns have been identified in other stream systems throughout western New South Wales (e.g., Fanning 1999, Pickard 1994).

Major changes in channel migration patterns have also occurred during the last 150 years. Pickard (1994) identified channel migration rates as high as one to two metres per year near White Cliffs, northeast of Broken Hill for the period since pastoral landuse commenced. However, unlike Wasson and Galloway (1986), he was reluctant to attribute these solely to the impact of pastoral landuse, pointing out that a similar rate of channel migration occurred prior to the arrival of sheep in western New South Wales. Instead, Pickard (1994:281) suggested that the current condition of creeks might represent the most recent morphological adjustment to changing conditions of flow and sediment yield.

Increased incidence of flooding, channel migration and scouring have implications for the preservation and visibility of the surface archaeological record (Fanning 1999:191). Along many areas adjacent to creeks, sediments deposited during flooding may have obscured much of the surface. Elsewhere erosion has exposed surface archaeological distributions of stone artefacts and heat retainer hearths across large areas of the landscape. Both these processes can also result in the scouring and destruction of archaeological deposits.

Palaeoenvironmental Record of Western New South Wales

There is limited information for the palaeoenvironmental record of western New South Wales. Available data points to variable environmental conditions from the mid Holocene onwards. Pollen cores from two small lakes near the junction of the Murray and Darling Rivers indicate a decrease in belah (*Casuarina pauper*) woodland after 4,500 years BP and an increase in chenopod shrublands (Cupper

1998:29). Cupper (1998) suggests that this transformation may be due to either climate change or an alteration of the fire regime. Increased soil salinization is one possible environmental change that may explain the pollen record.

The evidence for vegetation and environmental change combined with the geomorphic record indicate that the environment of western New South Wales is far from stable. This is further illustrated during the recent past where short-term rainfall variability is viewed as a contributing factor to geomorphic change. In conjunction with pastoral landuse, increased sediment yields in creeks after 1900 is attributed to a period of decreased rainfall resulting in erosion between 1910 and 1945 (Pickard 1994).

Summary

The present landscape of western New South Wales is an ideal location to investigate the surface archaeological record of the arid zone. The last 150 years of European land use has resulted in the exposure of spatially extensive archaeological distributions across large areas of the landscape. Although widely visible, many areas of this record are being destroyed through current land use and geomorphic processes, but little is understood about the spatial and temporal scales at which this is occurring. The destruction of the archaeological record is of concern, but many of the processes that are uncovering and ultimately destroying surface deposits also operated during the past. Therefore, an understanding of landscape history is a prerequisite for the interpretation of the archaeological record within a behavioural framework; especially those based upon comparisons between multiple locations.

The physical landscape of western New South Wales is dynamic. The agents of geomorphic change are complex. At the scale of a year or even a decade the landscape may appear to be in a state of equilibrium. However, within the broader scales represented in the accumulation of the archaeological record the development of the landscape is characterised by episodic non-equilibrium. This refers to periods of stability and sediment accumulation interspersed with periods of erosion that remove the record of the preceding accumulation phase (including archaeological materials). The implications of episodic non-equilibrium for the formation of archaeological deposits are just beginning to be explored (Fanning 2002, Fanning and Holdaway 2001a). At a general level, episodic non-equilibrium combined with occupational chronologies e.g., those established at Stud Creek demonstrate the complex temporal character of the surface archaeological record.

The Pine Point-Langwell Study Area

Landforms primarily reflect the geomorphic history of the landscape. However, the presence and location of different landforms is important in an archaeological context, with many containing archaeological deposits. The formation history of landforms also provides clues to the role of geomorphic processes in the formation or destruction of archaeological deposits and as demonstrated in previous archaeological studies (Bowler 1998, Fanning and Holdaway 2001a) is useful for establishing a relative chronological context for the archaeological record. The following sections describe the landscape history of the Pine Point-Langwell study area in terms of land systems, rainfall trends and the presence and distribution of different landform types.

General Description

Quaternary sediments of the Murray Basin Plain are ubiquitous on Pine Point and Langwell Stations. The western and northern boundaries of Pine Point Station also contain extensive residual and colluvial deposits of poorly sorted sand and gravel that surrounds isolated outcrops of Tertiary silcrete and Proterozoic gneiss of the Willyama Complex. The residual and colluvial deposits are bordered by extensive flat to gently undulating plains of red and brown clayey sand extending to the south. The floodplain and outwash deposits of Pine Creek dissect these plains. The study area is characterised by low relief, with elevations ranging between 100 and 195 metres above sea level. The highest elevation in the study area is Enmore Hill in the west at 195 metres above sea level. The fieldwork component of this study focused on the surface archaeological record found on landforms associated with the major fluvial system of the study area i.e. the Rantyga – Pine Creek confluence.

Land Systems

Christian and Stewart (1953:21) define land systems as an area, or areas, with a recurring pattern of topography, soils and vegetation. The Soil Conservation Service of New South Wales (1991), now part of the Department of Land and Water Conservation has mapped them over the whole of the Western Division of New South Wales and these provide a useful method of summarising the present environmental characteristics of the study area. The Pine Point-Langwell study area contains four different land systems (Figure 4.1):

1. Katalpa (KT)/Oakvale (OK) – level to undulating plains of the Barrier Range foothills west of the Silver City Highway and northeast of Rantyga Creek. Weathered pre-Mesozoic bedrock forms rises between nine and 30 metres in relief with minor areas of silcrete, ferricrete and other Tertiary sediments. Outcrops of silcrete are found on a series of low rises northeast of the Rantyga Creek-Pine Creek confluence. Quartz and silcrete gibber pavement occurs throughout. Low Chenopod shrubland of bladder saltbush (*Atriplex vesicaria*), pearl bluebush (*Maireana sedifolia*), black bluebush (*Maireana pyramidata*), ruby saltbush (*Enchylaena tomentosa*) are the predominant vegetation types. Trees restricted to scattered stands of mulga (*Acacia aneura*) and belah (*Casuarina pauper*) represent remnants of a previously more widespread community across western New South Wales (Fanning 1994:176).

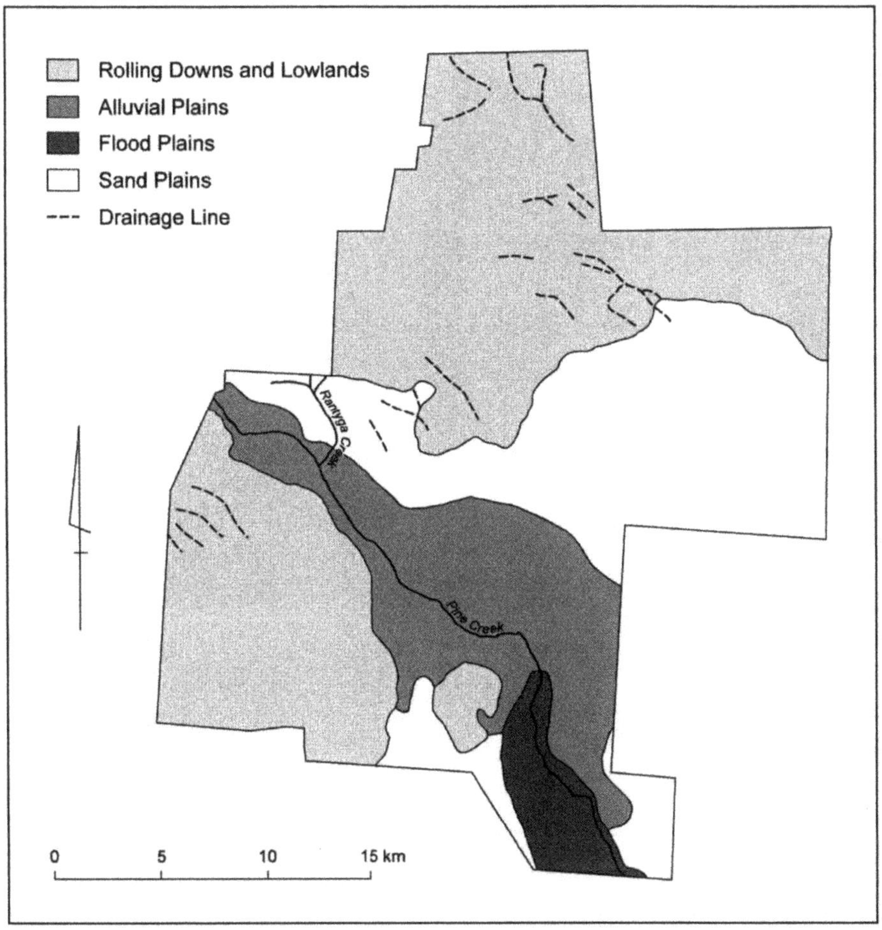

Figure 4.1 Land systems of the Pine Point-Langwell study area.

2. Kars (KZ) – flat to gently undulating sandplain of red and brown clayey sand, loam and lateritic soils characterised by low rises and widely spaced choked watercourses and rounded interfluves. Minor claypans occur across the plain. Channel and overbank sediments of Pine and Rantyga Creeks, consisting of sand, silt, clay and minor gravels dissect the plain. Clumped stands of black bluebush (*Maireana pyramidata*), goathead burr (*Sclerolaena bicornis*), silky copperburr (*Sclerolaena eriacantha*), and streaked poverty bush (*Sclerolaena tricuspis*) are widely distributed. Prickly wattle (*Acacia victoriae*), belah (*Casuarina pauper*), nelia (*Acacaia loderi*), mulga (*Acacia aneura*) and rosewood (*Heterodendrum oleifolium*) form clumped stands. Curly mitchell grass (*Astrebla lappacea*) and rough speargrass (*Stipa scabra*) are found along the margins of Rantyga Creek.
3. Conservation (CN)/Fowlers (FL) – Narrow channel and grey clay floodplain soils of Pine Creek bordered by a scalded alluvial plain and backplain of reddish-brown sandy and loamy textured soils with occasional bare gravelly rises. Stands of river red gum (*Eucalyptus camaldulensis*) are found in and immediately adjacent to Pine Creek with sparse saltbush (*Atriplex vesicaria*) occurring across the backplain.
4. Caloola (CL) – Terminal floodplains of Pine Creek consisting of deep self mulching brown and grey clays covered by saltbush (*Atriplex vesicaria*). River red gum (*Eucalyptus camaldulensis*) is found in and immediately adjacent to Pine Creek. This land system contains few scalds, and abundant overbank alluvial sediment.

Recent Rainfall Trends

In common with other areas of the arid zone, the study area presently experiences a low and variable rainfall. Table 4.2 contains the monthly rainfall totals for Pine Point station from 1991 to 2002. Although the rainfall distribution for the Pine Point area is described as uniform, there is considerable variability in the amount of rainfall received per month from year to year. The dry conditions contribute significantly to the extensive erosion and surface exposure across the study area. Vegetation quickly dies during dry periods and soil structure breaks down, increasing the rate at which it is eroded by alluvial and aeolian processes.

The variable character of rainfall means that it is presently difficult to predict when and where natural sources of water will be available across the properties. Part of this reflects changes to drainage during the period of pastoral landuse, but some of the unpredictability reflects general environ-

Table 4.2 Monthly rainfall total (mm) for Pine Creek station from 1991 to 2002. These indicate considerable variability in the amount of rainfall received each month and each year. The total average rainfall over this period is 226.4±124.9 millimetres.

Year	Jan.	Feb.	March	April	May	June	July	Aug.	Sept.	Oct.	Nov.	Dec.
1991	20.5	0	0	11	0	27	19.8	1.5	2.5	0	3.3	0
1992	0	18.5	6.5	0	7.5	16.5	0	20	39.3	50	54.8	303.5
1993	85	15.5	0	0	55.5	0	39	1	44	43.5	3.5	91.5
1994	0	106.3	8.5	0	0	22.5	0	0	30	11	0	0
1995	31	22.5	0	0	42.5	0	8	6	30	45.5	13.8	12.5
1996	16	18	100	0	0	26.3	48.3	27	49	22.5	10.3	17.5
1997	12.8	36.8	.5	0	23.5	0	0	12.5	33	30	16	21.5
1998	68.8	11.5	0	3	0	8.5	21.3	12	48	6	0	0
1999	8	16.5	43.3	0	2	5.5	35	2	20.5	17.3	28.5	13.3
2000	0	36.5	0	44.8	21.8	4.5	24.8	17.5	17	25	12	7.5
2001	30.3	12	0	0	5.8	27.5	18	8.5	32.5	22.3	0	1.8
2002	0	3	2	11	13.5	4.5	6	1.8	4	0	11.5	26.5

mental conditions. Box swamps are a good example of this unpredictability. On Pine Point Station box swamps have not contained a significant amount of water since the large rains during December 1992 (Ron Harvy Pine Point Station personal communication 2001). This level of unpredictability and the arid environment would be major considerations for hunter-gather people occupying the landscape. While these recent rainfall values and subsequent environmental characteristics are not a mirror of those during the past, they do highlight some of the general environmental conditions that would characterise the landscape.

Landscape Features and Landforms of Pine Point and Langwell Stations – Their History and Archaeological Prospects

Creeks

The study area contains a number of creeks of varying sizes. All of these are currently ephemeral and only contain water following heavy rainfall within their catchments. Projecting the present hydrological scheme onto the past is problematic, because many changes in drainage patterns and flow rates have resulted from activities associated with pastoral landuse within the wider western New South Wales region (Pickard 1994).

Pine Creek is the largest creek in the study area and has a catchment area of 1785 square kilometres. Originating in the Barrier Range to the west of Broken Hill, it flows south, terminating at a playa lake on Kudgee Station, approximately 30 kilometres downstream of Langwell station. The creek is a meandering distributary system that transports a large sediment load derived from its upland catchment in the Barrier Range. These sediments are deposited in and adjacent to the current channel.

In common with many other creeks in western New South Wales, the bed of Pine Creek is choked with sediment and vegetation, leading to a reduction in water carrying capacity and an increase in stream width relative to depth. Many of these changes have occurred within the period of living memory. On Langwell Station, for example, large numbers of *Eucalyptus camaldulensis* only began to colonise the bed and margins of Pine Creek during the 1940s and waterholes filled with sand.

Originating in the Barrier Range south of Broken Hill, Rantyga Creek is the second largest creek in the study area with a much smaller catchment area than Pine Creek of 203 square kilometres. Rantyga Creek is a tributary of Pine Creek, although substantial amounts of its flow are presently diverted to Rantyga Tank, a livestock watering point. Similar processes to those occurring along Pine Creek are operating along Rantyga Creek. There are numerous smaller order creeks and drainage lines across the two stations. These often have uncoordinated drainage patterns and terminate within areas of internal drainage, e.g., playa lakes. Surface archaeological deposits are abundant along the margins of both major creeks and are also found along drainage lines, particularly in the west of the study area.

Playa Lakes

Playa lakes are flooded less than a quarter of the time (Mabbutt 1986:200). The numerous semi-permanent and ephemeral lakes throughout western New South Wales fit this description. Typically, these form through the termination of stream flows at inland basins away from the Darling River, reflecting the disorganised nature of drainage systems in arid environments. Lakes along the Darling River and Anabranch floodplains are different; they receive water during periods of high river discharge (C. Webb 1996).

Pine Point and Langwell Stations contain a number of playa lakes of various sizes, but these are much smaller than the lakes in the Menindee and Willandra systems. They receive water from small internal drainage basins, usually covering areas less than several square kilometres and contain water at irregular intervals, only filling following periods of high rainfall. Like larger playa in other areas of the arid zone, the lakes on Pine Point and Langwell Stations have source bordering lunettes and sand dunes along their eastern margins. These have formed through similar processes to those described below for source bordering dunes. Vegetation along the margins of playa consists of black box (*Eucalyptus largiflorens*) with an understory of various salt bush (*Atriplex spp.*) and blue bush (*Maireana spp.*) species. Surface scatters of stone artefacts and heat retainer hearths are found along the margins of playas.

Source Bordering Dunes

Source bordering dunes are aeolian deposits derived from the deflation of local alluvial sediments during periods of seasonally low flow associated with increased aridity (Bowler 1986:126-127). They are most often found on the eastern and northeastern margin of prior streams on the Riverine Plain of the Murray-Darling Basin, reflecting the westerly direction of prevailing winds. The most recent period of source bordering dune formation in western New South Wales is dated to 15,000 to 13,000 years BP (Bowler 1986:127). Source bordering dunes occur along the Great Darling Anabranch, but are largely absent from the present Darling River channel (Balme and Hope 1990), which became active around 11,000 years BP. Studies in the Riverina area of New South Wales also associate source bordering dune formation with palaeo-channels, with the most recent period of dune formation dated to 12,000 years BP (Page *et al.* 1996). Bowler (1986) attributes the lack of source bordering dune formation along present channels to the development of riparian woodlands, which provide a protective vegetation canopy preventing the aeolian deflation of stream sediments.

In the study area, source-bordering dunes occur along the eastern and northeastern sides of Pine and Rantyga Creeks, and associated palaeo-channels. These consist of irregular crests separated by areas of scalding and deflation, with wind erosion resulting in the formation of blow-outs. Archaeological studies of source bordering dunes along the Darling River and the Great Darling Anabranch indicate that these contain burials, distributions of stone artefacts and heat retainer hearths (Balme and Hope 1990, Littleton 2000). Along Pine Creek, extensive stone artefact distributions and hearths are found on scalds and deflated areas within source bordering dunes.

Floodplains and Depositional Tributary Fans

Floodplains are flat areas adjacent to watercourses subject to inundation during floods (A. Brown 1997). They consist of sediments deposited during floods and provide a sedimentary record of floodplain evolution. In most cases surface archaeological deposits on floodplains will be recent in age (C. Webb 1996).

Pine Creek has a series of floodplains, extending up to several hundred metres from the present channel. The deposition of sediment on these has obscured much of the surface archaeological record in the CL and CN/FL land systems that encompass Pine Creek. Exposure in the CL land system is confined to a series of small scalds near the Pine Creek Bridge on the Silver City Highway. The presence of flood debris around the base of trees indicates that floodwaters have flowed over these exposures and hence there is a possibility of fluvial disturbance to artefact distributions here. Extensive areas of exposure occur in the CN/FL land system near and downstream from the confluence of Pine and Rantyga Creeks, but these are not widely distributed and are mainly confined to terraces rather than on the floodplain itself.

A major depositional tributary fan occurs at the confluence of Rantyga Creek and Pine Creek. This is characterised by the deposition of fine sediments derived from the Rantyga Creek catchment. Archaeological visibility on this landform is low.

Palaeo-Channels

Otherwise known as billabongs or oxbow lakes, palaeo-channels are remnant sections of abandoned streams found on prior or existing floodplains. Palaeo-channels form as a result of stream migration and are associated with meandering systems. This occurs when the deposition of sediments forms a plug that separates a meander loop from the main channel or when water velocity is sufficient, particularly during floods, to cut-off a meander loop (A. Brown 1997).

In the study area, numerous palaeo-channels occur along the northern side of Pine Creek. These are curved in shape and partially infilled with sediments deposited during floods. Archaeological materials occur on terrace and dune surfaces adjacent to palaeo-channels; some of these deposits may be old relative to deposits along present channels.

Terraces

Terraces form through alluvial processes operating within stream valleys (A. Brown 1997, Waters 1992). Characteristic features of a terrace include the tread (a more or less horizontal surface) and a scarp separating the terrace from the floodplain or terrace below (Waters 1992). There are two types of terrace, erosional and depositional. Erosional terraces are formed by stream incision into the floodplain (A. Brown 1997). Treads of erosional terraces are pre-existing older surfaces exposed through erosion (Waters 1992). Depositional terraces are formed by the deposition of sediment and the tread represents a former floodplain now abandoned through incision (Waters 1992).

Establishing the chronology of terrace development can be complex, and all of the sediments that constitute a single terrace are not necessarily the same age. Processes operating within one area of a fluvial valley may remove or cover different sections of the depositional record. However, a time window within which an individual terrace developed can be established by identifying the periods of fluvial incision that separates it from other terraces (A. Brown 1997:36-37).

In general, terraces are a chronological set of landforms that are younger nearer to the floodplain. Higher terraces are usually older and have therefore been available for human occupation for longer than lower terraces (Waters 1992:152). The temporal scale over which artefacts have accumulated on higher terraces is likely to be greater than that for lower terraces (Waters 1992:152).

Within the study area terraces occur along Pine Creek and to a lesser extent Rantyga Creek. Stone artefact scatters and hearths are common on these terraces.

Scalds

Artefact exposure and visibility is highest on scalded surfaces in the study area. As previously described, scalds are not landforms, but rather occur on a range of different landforms. Extensive scalding occurs along both Pine and Rantyga Creeks and on the surrounding plains, although the amount of scalding, particularly along Pine Creek has decreased since the 1940s (D. Harrison, Langwell Station, personal communication 2001).

Discussion

The different landforms present in the Pine Point-Langwell study area potentially document assemblage accumulation over varied amounts of time. The characteristics of these landforms provide clues as to the relative chronological context of archaeological distributions. For example, assemblages from source bordering dunes are likely to represent accumulation over a greater period than scalded surfaces near the present floodplain of Pine Creek. This knowledge combined with the results of previous geoarchaeological research in western New South Wales (e.g., Bowler and Price 1998, Holdaway *et al.* 2002a) cautions against assuming that widely dispersed archaeological distributions on varied landforms are contemporaneous. This highlights the importance of establishing chronological contexts.

The present study is following on the footsteps of the last 30 or more years of archaeological research in western New South Wales, where considerable efforts have been made to establish chronological contexts for archaeological materials (e.g., Allen 1972, Balme and Hope 1990, Bowler and Price 1998, Holdaway *et al.* 2002a, Martin and O'Donnell 1995). However, it seems that the important contribution of this work has largely been missed by a number of archaeologists working in other parts of the arid zone. This is a major factor contributing to the development of large-scale synchronic settlement system models. In western New South Wales the role of environmental variability has a long history of being recognised as a factor that restricts archaeologists from developing models of synchronic settlement systems (e.g., Allen 1990, 1998, Bowler and Price 1998).

Table 4.3 Potential chronological contexts of landforms in the Pine Point-Langwell study area.

Landform	Location in Study Area	Description	Chronological Context
Modern Creek Channels	Pine Creek, Rantyga Creek	Meandering channels	Deposits immediately adjacent are likely to be recent in age
Terraces	Alluvial valleys along Pine Creek and Rantyga Creek	Stepped land surfaces formed through down cutting of alluvial channel	Establishing the age of scalded surfaces on terraces can be difficult (e.g., Webb 1996). Recent studies (e.g., Holdaway *et. al* 2002a, 2002b) demonstrate that these surfaces may represent accumulation from the mid-Holocene onwards.
Source Bordering Dunes	Pine Creek	Sand hills adjacent the present channel and prior streams	Terminal Pleistocene, potentially document old deposits
Palaeo-Channels	Pine Creek	Located up to 500 metres from the present channel	Potentially old deposits
Floodplains and Depositional Tributary Fans	Pine Creek flood out, Pine-Rantyga Creek confluence	Areas that receive a regular sediment input from alluvial systems	Recent
Playa Lakes	Throughout	Small internal drainage basins	Unknown

The potential chronological contexts of the landforms discussed in the previous section are outlined in Table 4.3. Clearly the chronological record of accumulation is likely to vary considerably across the study area. The present study examines the archaeological record from scalded surfaces upon terraces near the confluence of Pine and Rantyga Creeks. Although this study is restricted to deposits from one single landform type it is possible that between these there is considerable chronological variability in the timing and tempo of archaeological accumulation. It is also important not to underestimate the significance of changes in hydrology across the study area. While it is true that many of these have occurred during the pastoral landuse period, many of the different landforms described in Table 4.3 are the direct result of changing alluvial contexts. Importantly the present landscape is not a direct analogue for that of the past. Some elements are similar; for example the distribution of stone materials, but others are not. Significant changes have occurred to the channel of Pine Creek, particularly in terms of increased sediment load.

Summary

Unlike the more conventional environmental descriptions associated with Australian arid zone settlement-subsistence studies, the present chapter has not attempted to model predicted functional associations between the environment and assemblage composition. Instead, this chapter focused on the role of environmental attributes in the formation of the archaeological record. Central to this is the acknowledgement that the geomorphological history of landscape evolution is an important variable, influencing the macro scale spatial and temporal distribution of the archaeological record. The limited amount of previous archaeological research in the region provides a general archaeological context. This work highlights the existence of surface archaeological distributions across a range of landform contexts, but detailed summaries of assemblage composition are lacking, as are radiocarbon determinations from heat retainer hearths.

A chronological model for the investigation of deflated surface archaeological deposits in western New South Wales was established in the previous three chapters. An important element of this model is an understanding of the geomorphic history of the land surfaces upon which archaeological materials are distributed. It is from this context that the landscape history of the wider western New South Wales region and that of the Pine Point–Langwell study area are discussed in detail. A range of land surfaces exists within the study area. The geomorphic context of these surfaces provides a relative, although very coarse, chronological control for the accumulation of archaeological materials. The following chapter discusses the archaeological survey and recording methodology.

Chapter Five: Archaeological Survey and Recording Methods

Introduction

In any archaeological study it is important to describe in detail the methodological techniques applied to data collection and analysis. The structure of data collection strategies has implications for the types of interpretations derived from archaeological materials. Therefore, it is important to clearly specify the method by which different types of information are collected. This ranges from clear definitions of individual artefact types through to detailed descriptions of the recording protocol employed.

Three elements of the surface archaeological record were investigated in this study; surface stone artefact distributions, heat retainer hearths and stone sources for the production of artefacts. A computerised primary data collection system was designed to maximise accuracy, precision and efficiency, and to integrate lithic, spatial and geomorphological data into a single database for analysis (see Dibble and McPherron 1988). In addition, the variable definition and recording protocol was designed to follow as closely as practically possible that of WNSAWP, thus allowing complete data exchange and integration between the two.

Fieldwork Aims

The fieldwork methods were designed to meet several specific aims related to the overall goals of this research, namely to investigate place use histories and diachronic variability between stone artefact assemblages from deflated surface deposits. These aims were:

- To establish occupational chronologies through obtaining radiocarbon determinations from partially excavated heat retainer hearths and, observations of the geomorphic context of four stone artefact distributions from the Pine Point-Langwell study area.
- To record stone artefact assemblages of sufficient size (>3000 pieces) so as to minimise the influence of outliers in analyses of central tendency that arise from insufficient sample size and to permit a comprehensive range of technological analyses within and between assemblages.
- To record the spatial location of stone artefacts over areas sufficiently large (>5,000 square metres) to identify patterns in artefact distribution.
- To record the micro-geomorphic context of artefact distribution to allow an investigation of the influence of geomorphic processes on artefact visibility and the spatial structure of artefact distribution.

Consultation

The consultation process for the study involved discussions with members of the local Aboriginal community and pastoralists within and around the study area.

Consultation with the Broken Hill Local Aboriginal Land Council (BHLALC)

Initial approaches regarding the feasibility of the project were made to Mr. Badger Bates of the New South Wales National Parks and Wildlife Service (NSWNPWS), Broken Hill during April 2000. Mr. Bates suggested several possible locations where archaeological fieldwork could be undertaken and facilitated further discussions with the Broken Hill Local Aboriginal Land Council (BHLALC). Following selection of the field area on Pine Point and Langwell Stations, discussions regarding the nature of the fieldwork commenced with Mr. Raymond O'Donnell (Aboriginal Sites Representative, BHLALC) in June 2000. Mr. O'Donnell participated in the WNSWAP fieldwork at Fowlers Gap Station between 1999 and 2001 and therefore had first hand experience of the proposed research activities. Additionally, Mr. O'Donnell provided written support on behalf of the Broken Hill Local Aboriginal Land Council Board for a permit to the New South Wales National Parks and Wildlife Service for the excavation of 30 heat retainer hearths as part of the project. Regular information regarding the progress of the project was been communicated to the Broken Hill Local Aboriginal Land Council through discussions with, and letters to Mr. O'Donnell.

Consultation with Local Landholders

Access to the field area was negotiated with the Harvy Family of Pine Point Station and the Harrison Family of Langwell Station. Both families consented to the proposed research being undertaken on their properties, with the Harvy family providing generous access to accommodation and other facilities.

Selection of the Study Area

Selection of the study area was undertaken during a one week visit to Broken Hill in April 2000. The aim of the visit was to identify an area outside of the Barrier Range where the spatial and compositional characteristics of surface stone artefact scatters could be studied. Between 1995-1998 and 1999-2001 WNSWAP investigated the surface archaeological record at Mt Wood in the Grey Range and at Fowlers Gap in the northern Barrier Range respectively. The initial reason for selecting this study area on the plains was to obtain data for comparison with Mount Wood and Fowlers Gap, in an area outside of the ranges.

After the inspection of several areas to the east and west of the Barrier Range, Pine Point and Langwell Stations to the south were selected. Although in a different geomorphic setting to the ranges, the location and distribution of archaeological exposure suggested that similar geomorphic processes had led to the formation of deflated surface distributions to other WNSWAP survey locations. In addition, the occurrence of silcrete and quartz sources in the study area minimised differences in raw material access between the Pine Point-Langwell study area and the other study areas situated in the ranges.

Fieldwork

Four seasons of fieldwork were conducted between January 2001 and September 2002. An initial two-week field season during January 2001 investigated the distribution of archaeological material across the study area. A reconnaissance survey strategy was designed and implemented, and a method for recording the spatial and physical attributes of stone artefacts was formulated and tested during this period. A six week field season between July and August 2001 focused on the recording of artefact distributions at sampling areas along Rantyga Creek and Pine Creek. Hearths were also recorded and excavated in several sample areas. Between January and February 2002 the recording of artefact distributions along Pine Creek was completed and artefact recording was commenced at the KZ2 sampling area along Rantyga Creek. The nature and distribution of potential stone sources suitable for the production of stone artefacts throughout the study area was recorded and excavation of more hearths along Rantyga and Pine Creeks was also undertaken. The final field season was conducted between late May and September 2002. This involved completing the recording at KZ2 and additional excavation of hearths along Pine and Rantyga Creeks. Artefact recording was undertaken at silcrete quarry 1 (SQ1) in the low hills north of Pine Creek.

Field Survey and Site Selection

The two stations cover a combined area of 73,000 hectares, and contain a range of different landforms upon which surface archaeological deposits are widely distributed, as described in Chapter Four. To efficiently identify areas where the aims of the study could be addressed, a multi-stage reconnaissance survey strategy was undertaken:

1. The 1:250,000 Menindee land systems map (Soil Conservation Service of New South Wales 1982) was used to identify the four land systems found in the study area. Land systems have been widely used by Australian archaeologists in stratified surveys (Holdaway *et al.* 2002b, Robins 1993, Thorley 1998b, Thorley *et al.* 1994). They are useful for assessing conditions of exposure across a landscape because the geomorphic processes that influence artefact exposure and visibility vary more between land systems than they do within them (Holdaway *et al.* 2002b). The four land systems formed the basic sampling strata for subsequent surveys.
2. Aerial photographs were used to identify scalding and other areas of high surface exposure where the archaeological record is expected to be highly visible. These appear as distinctive bare orange areas on the photographs.
3. These areas were located on 1:100,000 topographic maps and ground checked with a handheld 12-channel GPS (global positioning system). Once accurately located, these areas were intensively foot surveyed to assess their archaeological potential in terms of the degree and condition of exposure, geomorphic context, the degree and condition of artefact distribution and the presence of hearths likely to contain charcoal.

Sample Area Selection and Description

The selection of the sample areas was one of the most important facets of the fieldwork programme.

Results of the Reconnaissance Survey

The reconnaissance survey indicated that spatially extensive surface archaeological deposits are widely distributed across the two stations in a range of landform contexts. An archaeological assessment of individual landforms within each of the land systems was made (Table 5.1). As can be seen from this table, the potential of landforms within different land systems for investigation varied considerably. For example, artefact distributions on the present floodplain of Pine Creek, especially on Pine Point Station, exhibit evidence of spatial re-organisation e.g., linear size sorting in a downslope direction in rills. In other areas, for instance on the northern valley slopes of Pine Creek near the Pine Point-Langwell boundary fence, a steep gradient and associated rilling suggests the potential for down slope movement of artefacts. Despite some surfaces not being suitable for artefact recording, the frequency and extent of remaining artefact distributions across the study area made it impossible to incorporate them all within the design of the present study. Instead, it was decided that the aims of the study could be addressed through the detailed investigation of surface deposits within the vicinity of the Rantyga Creek-Pine Creek confluence.

This area was selected because it was felt that a large sample of stone artefacts could be obtained from the numerous and spatially extensive scalded surfaces upon the terrace running along Pine Creek. In addition, an occupational chronology could be established through the dating of charcoal bearing heat retainer hearths on this surface. The Rantyga Creek area was selected for similar reasons. Extensive areas of scalding on the distal floodplain of Rantyga Creek contain distributions of stone artefacts and heat retainer hearths with indications of charcoal. Although charcoal-bearing hearths were not as numerous as those on the Pine Creek terrace, it seemed likely that a number of radiocarbon determinations could be obtained from hearths along this alluvial system. Sandy rises also contain areas of exposure with varying densities of stone artefacts.

Table 5.1 Archaeological assessment of individual landforms within each land system for the Pine Point-Langwell study area.

Land System	Landform	Archaeological Assessment
Conservation-Fowlers (CN/FL)	Source Bordering Dune	Isolated exposures in blow-outs with stone artefacts and occasional heat retainer hearths
	Terrace	Numerous scalds with extensive distributions of stone artefacts and heat retainer hearths
	Floodplain	Sediment deposition has obscured/and or removed artefact distributions
Caloola (CL)	Terminal Floodplain	Sediment deposition has obscured/and or removed artefact distributions
Katalpa-Oakvale (KT/OK)	Low Rise	Soft scald exposures with sparse artefact distributions, worked silcrete outcrops with extensive artefact distributions and no hearths
	Drainage Line	Numerous scalds with extensive distributions of stone artefacts and heat retainer hearths with little prospect of retaining charcoal
Kars (KZ)	Distal Floodplain	Numerous scalds with extensive distributions of stone artefacts and heat retainer hearths
	Proximal Floodplain	Sediment deposition has obscured/and or removed artefact distributions
	Sandy Rise	Punctuated surface exposure with artefact distributions and artefacts and hearths eroding from sediment
	Slope	Soft scald exposures with sparse artefact distributions
	Box Swamp	Punctuated surface exposure on low sandy dunes on swamp margins with distributions of stone artefacts and heat retainer hearths containing charcoal

Surfaces were not selected from the CL and KT/OK land systems for several reasons. First, the study primarily aimed to understand the nature of assemblage formation through the intensive investigation of place use histories within a temporal framework. This is not possible for the CL and KT/OK land systems due to the absence of well preserved hearths. Second, with the limited resources available for the fieldwork programme, this could only be undertaken at a small number of places. Third, the CL land system contained a very small number of surfaces where artefact distributions were visible and hearths in the KT/OK land system showed no signs of retaining charcoal for dating. The distribution of potential sources of stone for the production of artefacts is the only feature of these land systems incorporated into the study.

Sampling Areas

As established in Chapters One and Two, many of the problems associated with settlement pattern studies arise from the coarse spatial and temporal scale at which artefact distributions are recorded. Some studies provide only cursory observations of the temporal context of archaeological accumulations. For this reason it was decided in the present study to confine recording to a small number of surfaces in the Rantyga Creek-Pine Creek confluence. To undertake a settlement pattern study across the whole range of landforms and ecological contexts represented in the Pine Point and Langwell study area would require that the temporal character of the record to either be ignored or studied extensively. The financial resources were not available to obtain radiocarbon determinations from all of the potential recording locations in a wider scale study. Even when the focus of the study is limited to the area around the confluence, there are still many possible surfaces in different micro-environmental settings that could be investigated.

With these considerations in mind, a number of potential sampling areas were identified in each land system. Potential sampling areas were coded according to land system and sequentially numbered. Finally, four different sampling areas (two from the CN land system and two from the KZ land system) were selected for detailed recording (Figure 5.1). The rationale for selecting these was twofold. Firstly, each location contained extensive areas of high land surface exposure, resulting in relatively high visibility of deflated stone artefact distributions and heat retainer hearths. This permitted the recording of artefact distributions over areas of sufficient size for the investigation of spatial patterning and analysis of large numbers of artefacts. Secondly, the range of landforms upon which artefacts are distributed provided relative estimates of the maximum antiquity of the archaeological surfaces and in-turn the relative amount of time represented in artefact accumulation. In addition, the presence of charcoal bearing hearths permitted the establishment of occupational chronologies for the surface deposits.

The geomorphic features of the four locations were documented by surveying transects across each area parallel to the stream channel. Geomorphic units along each transect were identified, and the surface condition, dominant geomorphic processes and vegetation of each were described (Tables 5.2 to 5.5).

Table 5.2 Transect CN1A description.

Unit	Name	Description	Sediment/Surface	Vegetation
1	Active Channel Bank	Steeply sloping right bank of Pine Creek with abundant leaf litter and flood debris	Red silty loam	River red gum, occasional water bush and sparse onion weed
2	Active Channel Bed	Broad sand filled ephemeral channel bounded by stable banks with fringing gallery of river red gums	Loose pale red medium to coarse sand	River red gums germinating in channel bed
3	Active Channel Bank	Left bank of Pine Creek (As for Unit 1)	(As for Unit 1)	(As for Unit 1)
4	Proximal Flood Plain	Flood scoured bench with relatively dense river red gums	Mixed sandy and silty sediment	River red gums of various ages, the most mature being those lining the main channel
5	Terrace Slope	Gradual rise from proximal flood plain to terrace surface		
6	Terrace Surface – Scald	Broad, circular hard scald surface with stone artefacts and heat retainer hearths	Hard setting clay B horizon surface	None
7	Terrace Surface – Sediment Island	Patches of sandy sediment on highest remaining surface on terrace	Fine to coarse sand	Predominantly black bluebush, rare copperburrs and cannonballs

Table 5.3 Transect CN1B description.

Unit	Name	Description	Sediment/Surface	Vegetation
1	Active Channel Bank	Steeply sloping right bank of Pine Creek with abundant leaf litter and flood debris	Red silty loan	River red gum, occasional waterbush and sparse onion weed
2	Active Channel Bed	Channel very wide and divided into two by an elongate mud channel bar	Loose pale red medium to coarse sand	River red gums germinating in channel bed
3	Active Channel – Channel Bar	Mid-channel island, elongate in downstream direction	Mud drape over sandy sediment	None
4	Active Channel Bank	Left bank of Pine Creek (As for Unit 1)	(As for Unit 1)	(As for Unit 1)
5	Proximal Flood Plain	Flood scoured bench with relatively dense river red gums	Mixed sandy and silty sediment	River red gums of various ages
6	Terrace Slope	Gradual rise from proximal flood plain to terrace surface		
7	Terrace Surface – Scald	Broad, circular hard scald surface with stone artefacts and heat retainer hearths	Hard setting clay B horizon surface	None
8	Terrace Surface – Sediment Island	Patches of sandy sediment on highest remaining surface on terrace	Fine to coarse sand	Predominantly black bluebush, rare copperburrs and cannonballs

Table 5.4 Transect KZ1 description.

Unit	Name	Description	Sediment/Surface	Vegetation
1	Distal Floodplain – Sediment Island	Rain splash crust with veneer of lagged coarse sand, rare artefacts appearing at surface	Sandy/silty loam	Sparse clumps of copperburr, pearl bluebush and black bluebush
2	Distal Floodplain – Scald	Rain splash crust with sparse to dense veneer of lagged quartz and manganese granules, sparse distribution of artefacts as lag	Silty clay	None
3	Proximal Floodplain – Depositional	Undulating depositional surface, no visible Aboriginal artefacts	Sandy/silty loam	Sparse remnants of low shrubs, plus scattered belah, and clumps of neddlewood
4	Proximal Floodplain – Scoured	Undulating rainsplash surface with flood scours and hummocks sub-parallel with main channel	Loose sandy sediment, woody flood debris against upstream sides of vegetation hummocks	Circular mounds up to three metres with dense bluebush
5	Active Channel – Inset Bench	Erosional bench inside left bank of Rantyga Creek	Loose sandy/silty bright red sediment	Lemongrass, onionweed, sparse waterbush
6	Active Channel – Channel Bed	Wide (>15m), shallow trench, filled with sediment, meandering thalweg visible	Pale red sand, aeolian ripples visible in areas not disturbed by trampling	Noongoora burr, bathurst burr
7	Active Channel – Mid Channel Bar	Lozenge shaped low bar to right of main channel, separated from right bank by narrow channel about 2m wide	Loose medium to coarse sand and granules	Lemongrass, bathurst burr, noongoora burr and saffron thistle

Table 5.5 KZ2 Transect description.

Unit	Name	Description	Sediment/Surface	Vegetation
1	Distal Slope	Low angled slope, with sparse broken rain	Fine sandy to silty clay loam with abundant calcareous nodules	Groves of belah, black bluebush at two metre spacing
2	Proximal Floodplain	Narrow band of hummocky surface between slope and channel	Loose silty sediment, leaf litter and abundant woody flood debris mixed with sediment against upstream side of shrubs	Prickly wattle, African box thorn, occasional waterbush, copper burr, cannonballs
3	Active Channel – Channel Bed	Multiple distributary channels (3) forming a flood out upstream of belah woodland	Medium pale red loose sand with aeolian ripples heavily disturbed by sheep, kangaroo and emu	Scattered onion weed, saffron thistle, lemongrass on margins
4	Active Channel – Mid Channel Bar	Main channel has truncated to form two islands	Loose, unconsolidated silty sand veneer, abundant animal tracks	Variable densities of African box thorn, prickly wattle, onion weed, lemongrass is abundant, but discontinuous
5	Proximal Floodplain	Wide band of hummocky surface	Loose silty sediment, leaf litter and abundant woody flood debris mixed with sediment against upstream side of shrubs	Prickly wattle, african box thorn, abundant belah, occasional waterbush, copper burr, cannonballs
6	Distal Floodplain	Narrow depositional surface forming a truncation zone between the proximal floodplain and the sandy riser on the west side of Rantyga Creek	Broken rain splashed crust with coarse sandy lag	Occasional belah, black bluebush, and scattered copperburr
7	Sandy Riser	Broadly convex riser, with some scalds along crest, scalds contain abundant stone artefacts; some artefacts appearing on non-scalded surface, hearths appearing at surface	Coarse aeolian rippled sand lag, patchy leaf litter and abundant sheep tracks	Scattered copperburrs and rare black bluebush
8	Sandy Riser – Claypan	Circular depression west of crest of sandy riser, shallow drainage lines feed into claypan on northern edge	Rain splash crusted hard scald on rim, hexagonal cracked crusted hard surface in centre	Canegrass
9	Sandy Riser – Scald	Rain splashed crust with sparse to dense veneer of lagged quartz and manganese granules, variable distribution of stone artefacts and heat retainer hearths artefacts as lag	Silty clay	None
10	Sandy Riser – Eroded Flank	Densely rilled west facing slope; artefacts in rills are channelled in downslope direction	Rain splash crust, dispersed sand and granules of quartz and manganese	None
11	Flood Channel	Broad, shallow, elongate depression, with lateral gully headcuts into Unit 10, very active into Unit 12	Broken silty surface	Black bluebush, waterbush, prickly wattle, copperburr
12	Hill Slope	Low angled east facing slope, with patchy depositional surface interspersed with scalds, very sparse distribution of stone artefacts on scalds	Fine-grained silty surface on vegetation patches, hard rain splashed crust on scalds	Confined to depositional surfaces, black bluebush, unidentified herbs, copperburr

The characteristics of the four sampling areas are summarized below:

KZ1

The KZ1 sampling location is located on both sides of Rantyga Creek, approximately three kilometres upstream of its confluence with Pine Creek in the Kars land system. The channel of Rantyga Creek is a wide (>25 metres), shallow trench choked with sediment and covered by onion weed (*Asphodelus fistulosus*), bathurst burr (*Xanthium spinosum*), and waterbush (*Grevillea nematophylla*). The proximal floodplain is characterised by dense woody flood debris and flood scour holes immediately adjacent to the channel. This is bordered by a depositional proximal floodplain characterised by an undulating surface with sparse remnants of low shrubs, mainly hopbush (*Dodonaea viscosa*), and isolated belah (*Casuarina cristata*). The distal floodplain consists of longitudinal areas of scalding intersected by remnant islands of silty loam sediment. Vegetation consisting of pearl bluebush (*Maireana sedifolia*) and black bluebush (*Maireana pyramidata*) is confined to sediment islands.

Figure 5.1 Location of the sampling areas and the SQ1 silcrete quarry.

Sheetwash and aeolian deflation has exposed a sparse distribution of stone artefacts and heat retainer hearths on the distal floodplain. Artefacts and hearths are being exposed by erosion of remnant sediment islands. Artefact distributions were not visible on the proximal floodplain, where the deposition of overbank sediment has obscured any that may exist. Examination of the sediments along Rantyga Creek was made at a section on a tributary gully bordering the proximal margin of the distal floodplain on the left bank of the creek. Immediately below the surface to a depth of 22 centimetres is a massive, uniform, poorly sorted coarse sand to silt unit with no visible sedimentary structure representing post-European alluvium (PEM) (See Fanning 1999). In other areas of western New South Wales this sediment has buried and in many cases preserved archaeological distributions (Fanning 1999, 2002). In some locations the PEM is being stripped away by aeolian and alluvial erosion, exposing distributions of stone artefacts and heat retainer hearths.

Stone artefact distributions were recorded on a total of 21 scalded areas on both sides of Rantyga Creek over a two kilometre long section of the distal floodplain. A total of 3,419 stone artefacts were mapped and analysed over a sampling area of 11,795 square metres and 10 heat retainer hearths were recorded.

KZ2

The KZ2 sampling location is an 800 metre long area of discontinuous scalding situated on the right bank of Rantyga Creek, approximately 1.5 kilometres upstream of the confluence with Pine Creek in the Kars land system. Here Rantyga Creek consists of three distributary channels divided by a truncated mid channel bar. The right hand channel is the active channel, with the creek forming a minor floodout upstream of an area of belah (*Casuarina cristata*) woodland. The channels are choked with sediment and contain onion weed (*Asphodelus fistulosus*), saffron thistle (*Carthamus lanatus*) and lemongrass (*Cymbopogon ambiguus*). The left bank (eastern) side of the channel is bordered by a narrow proximal floodplain with a hummocky loose silty surface. Prickly wattle (*Acacia victoriae*), African box thorn (*Lycium ferocissimum*) and onionweed (*Asphodelus fistulosus*) are the dominant vegetation with minor occurrences of belah (*Casuarina cristata*) and waterbush (*Grevillea nematophylla*). A low angled distal slope borders the proximal floodplain, characterised by a fine sandy to silty loam surface covered with calcareous nodules. Artefacts are very rare on this surface. The right (western) side of the channel is bordered by a similar, but slightly wider proximal floodplain than the left side. A narrow distal floodplain forms a transitional depositional surface between the proximal floodplain and a sandy rise. A scoured surface indicates occasional flooding. Isolated belah (*Casuarina cristata*), black bluebush (*Maireana pyramidata*) and scattered copperburr (*Sclerolaena spp.*) are the dominant types of vegetation on the distal floodplain. This is bordered to the west by a broadly convex coarse sandy rise with scalding along the crest. A circular claypan depression covered by canegrass (*Eragrostis australasica*) occurs to the west of the crest of the sandy rise. Shallow drainage lines on the northern edge feed water into the claypan. Dense distributions of stone artefacts and heat retainer hearths are found on the scalds around the claypan. Artefacts and hearths are also visible, eroding from areas of remnant sediment on the sandy rise. The western side of the sandy rise is characterised by a densely rilled eroded flank. Artefacts in the rills are concentrated in the downslope direction.

The eroded flank borders a shallow elongate flood channel. The channel bed is vegetated with black bluebush (*Maireana pyramidata*), waterbush (*Grevillea nematophylla*), prickly wattle (*Acacia victoriae*) and copperburr (*Sclerolaena spp.*). Lateral gully headcuts are very active into the eroding flank on the east and the hillslope on the west. The low angled western hillslope is characterised by scald areas intersected by patchy depositional surfaces. Vegetation is confined to depositional surface and predominantly consists of black bluebush (*Maireana pyramidata*), copperburr (*Sclerolaena spp.*) and unidentifiable herbs. Artefacts are rare on the hillslope scalds.

A total of 11,192 artefacts and 26 heat retainer hearths were recorded on the scalds. Artefact distributions were mapped over an area of 7203 square metres. KZ2 is thought to be the archaeological distribution recorded as site 82 during the Mildura to Broken Hill powerline survey by McIntyre (1977). Site 82 is described as a "very extensive artifact scatter, cores, worked flakes, grinding slabs, scrapers, hammerstones, mixed stone types, burnt clay, quartz, hearths" (McIntyre 1977:37). The report recommends "absolutely no access" to the area encompassing the artefact distribution and the powerline is situated several hundred metres to the west of KZ2.

CN1

CN1 encompasses an extensive area of scalding located on top of a terrace on the right hand side of Pine Creek approximately 600 metres downstream of the confluence with Rantyga Creek in the Conservation land system. Two transects describe the CN1 sampling location. The first transect (CN1A) is located at the northwest edge of the area. Here Pine Creek is a broad sand filled ephemeral channel enclosed by steeply sloping banks with a fringing gallery of river red gum (*Eucalyptus camaldulensis*) and an understory of waterbush (*Grevillea nematophylla*) and sparse onion weed (*Asphodelus fistulosus*). Mature river red gums in the channel suggest minor lateral shift of the channel on the right hand side, whilst other areas of the channel are colonised by river red gum seedlings. The right hand bank of the creek is bordered by a scoured proximal floodplain with dense river red gums. There are major scour channels around the base of the trees and abundant woody flood debris on the upstream side of the trunks. A shallow scour channel running parallel to the main channel marks the distal boundary between the scoured floodplain and the terrace slope. The terrace slope forms a gradual rise from the proximal floodplain to the terrace surface. The slope is vegetated with abundant prickly wattle (*Acacia victoriae*), black bluebush (*Maireana pyramidata*) and copperburr (*Sclerolaena spp.*). Occasional patches of gravels (predominantly quartz) are visible on the slope. The top of the terrace is characterised by a circular hard scald surface with extensive distributions of stone artefacts and heat retainer hearths. Several sheep tracks cross the scald and there are numerous animal tracks in softer areas of the surface. Isolated remnant areas of subsoil appear as highpoints, as do occasional sediment islands.

The second transect (CN1B) is situated at the extreme southern end of the sampling area. The channel of Pine Creek is wider here and divided into two by an elongate mid-channel bar. Flood debris is concentrated on the upstream side of the trunks of river red gums in the channel. The channel abuts the terrace slope on the left hand side. Scouring at the base of the terrace slope has exposed a gravel bar of predominantly quartz clasts. Excavation of a bank section revealed that the gravel bar is outcropping gravely alluvium forming part of the terrace sediment. The lower terrace slope is dominated by prickly wattle (*Acacia victoriae*), whilst black bluebush (*Maireana pyramidata*) predominates on the upper slope. The terrace surface is the same as that described for the CN1A transect with the exception that vegetation is sparser on the sediment islands.

Stone artefacts and heat retainer hearths are predominantly found on the scalded surfaces of the terrace. Smaller numbers of artefacts and hearths can also be seen eroding from remnant sediment islands. Artefacts were rarely noted on the terrace slope and none were visible on the depositional surface of the proximal floodplain. A total of 8,788 artefacts were mapped and analysed over an area of 19,355 square metres at CN1. A total of 87 heat retainer hearths were mapped over a five kilometre long section running parallel to the left hand side of Pine Creek and encompassing both the CN1 and CN3 sampling areas.

CN3

The CN3 sampling location encompasses an 800 metre long area of discontinuous scalding situated on the left hand terrace running parallel to Pine Creek approximately one kilometre down stream of CN1 in the Conservation land system. The landform setting of CN3 is similar to that of CN1, therefore it was not necessary to describe a separate transect for the sampling area.

A total of 4,904 artefacts were mapped and analysed over an area of 12,813 square metres. In addition several major clusters of heat retainer hearths occurring within the sampling area were recorded during the hearth survey.

Spatial Mapping

Three phases of mapping was undertaken in each sampling area. The first involved recording the distribution of stone artefacts and micro-geomorphic surface condition. Grids, composed of one x one metre squares, were laid with the aid of tape and compass on a north - east axis. The maximum size for any individual grid was 2500 square metres. The number of grids laid in each sampling location depended upon the size and shape of the deflated exposures. Grids were laid until artefact distributions had been mapped over a minimum area of 5000 square metres. This had the benefit of increasing assemblage size, which was necessary at both the KZ1 and CN3 sampling areas because artefact densities were relatively low.

Artefact density and micro geomorphic context were important considerations governing the placement of sampling grids. Investigating the spatial character of artefact distributions is one of the main objectives of the spatial mapping programme. Rather than laying grids in the densest areas of artefact distribution (e.g., Veth 1993:30) grids were laid across distributions encompassing a range of different artefact densities. While this slowed the pace of artefact recording, it provided a comprehensive spatial record of artefact distribution patterns across each sampling area.

The first phase of the large grid mapping involved the selection of a scald upon which the grid/or consecutive grids would be laid. Each corner of the grid was marked with a large metal pin and tape measures were laid out between the pins. Tapes were then laid at one-metre intervals along a north-south axis beginning at zero. Tapes were laid continuously from the western edge across the grid until the eastern margin was encountered. This established a large grid with a series of one-metre squares. Every one-metre square in each of the sampling areas was assigned a unique and consecutive identification number. A nail with coloured tape was placed in the ground next to each individual artefact within the grid greater than 20-millimetres in maximum dimension. The 20-millimetre size cut-off for artefact recording followed previous work (Fanning and Holdaway 2001b, Schick 1987:96) that established a relationship between slope angle and artefact movement in surface distributions. Each artefact was assigned the square identification number of the square within which it was found. A DGPS (differential global positioning system) was used to obtain Australian Map Grid (AGM) co-ordinates for the four corners of each grid. These co-ordinates were used to generate grid coverages in ARCINFO GIS (Geographic Information System). These coverages were imported into the Arcview spatial analysis programme for further manipulation and analysis. The DGPS provided high precision coordinates. Differential GPS is far more accurate than conventional GPS, which may have errors of several hundreds of metres (Wheatley and Gillings 2002:73).

The second phase of mapping involved the construction a geomorphic base map for each sampling area. Initially the DGPS was used to record the location of major geomorphic features, i.e., scalds, creeks, non-lagged surfaces and rills. However, this was found to be time consuming and unnecessary because these features are clearly visible upon the aerial photographs. Instead, it was decided that the grid coverages could be overlaid on the appropriate aerial photograph. This involved rectifying each aerial photograph into the AGM co-ordinate system. To do this six control points with known AGM co-ordinates were selected on each of the aerial photographs. The aerial photographs were then geo-referenced in the Arcmap computer software package. The third phase involved mapping the location of hearths with the DGPS in each of the sampling areas.

Artefact Recording

All artefact recording was conducted in the field, with artefacts being identified, measured and replaced during a single recording session, thereby eliminating the need for collection. Artefact definitions follow Holdaway and Stern (2004). A range of attributes was recorded for each artefact. An attribute is a variable, and these summarise technological and typological characteristics (Richards and Ryan 1985). Not all attributes were recorded on all artefacts. When an attribute or measurement could not be recorded because the relevant feature was absent, a null value was entered. These attributes provide information on the nature of core reduction and tool production. Attribute selection was based on the WNSWAP stone artefact database, thus permitting the integration of artefact data from the Pine Point–Langwell study area with data from other WNSWAP project locations. The author conducted all artefact recording, and therefore inter-observer variability, described by Gnaden and Holdaway (2000) does not affect the quality of the dataset.

Artefact attributes were entered in to a HP 200LX palmtop computer running the Entrer Trois data entry program (McPherron and Holdaway 1996). This permitted the efficient recording of artefact attributes. In addition, the computerisation of recording limited the potential for errors because the data entry program prompts the user for input through the use of menus for nominal data (Holdaway *et al.* 1998). Metrical attributes were recorded with high precision electronic calipers. At the completion of each day's recording, artefact attribute data were downloaded and stored in the project database.

Micro-Geomorphic Surface Recording

Micro-geomorphic surface condition was recorded for each one-metre square within the sampling areas to assess the effect of differential visibility on artefact detection. The method is based on techniques applied in other areas of western New South Wales by Holdaway *et al.* (2002b). Eight different surface types were identified and allocated a separate numeric code. The square identification number and the numeric code of the surface condition of each square were entered into to a HP 200LX palmtop computer running the Entrer Trois data entry program (McPherron and Holdaway 1996). Because two or more surface conditions could be present in an individual square, the definition of surface condition was based on the type that accounted for the greatest proportion of the square. While this method is subjective in that some grids contained more than one surface type condition, the assessment throughout the project was consistent because the author was the sole assessor.

Hearth Recording and Excavation

Across western New South Wales, hearths commonly occur as varied concentrations of heat-fractured stone (Holdaway *et al.* 2002a, 2002b). These are often pedestalled a few centimetres above the surrounding ground surface as compact concentrations of heat retainers, but sometimes also occur as eroded and dispersed remains. Intensive pedestrian surveys for hearths were conducted within each of the four sampling areas. A total of 122 hearths were recorded during the survey. Each hearth was allocated the code of the survey area and a unique identification number. An aluminium tag containing this information was nailed into the ground near the hearth. A standard range of attributes about the surface structure, condition and excavation potential was recorded for each individual hearth and a digital image taken. The location of each hearth was recorded with the DGPS, which provided high precision AMG co-ordinates.

The New South Wales National Parks and Wildlife Service granted a permit for the excavation of 30 hearths in the study area. Hearths were chosen for excavation based on their preservation (primarily the likelihood that they would contain charcoal) and their archaeological context (only hearths within the vicinity of lithic recording areas were excavated). An image was taken and a plan drawn of each hearth prior to excavation.

The hearth excavation protocol used at Pine Point-Langwell follows that of Holdaway *et al.* (2002a). A plan of the intact surface was drawn, and the hearth was divided into four equal quadrants. Excavation commenced in one quadrant, with the removal of material stone by stone and layer by layer until the bottom of the fire pit was reached. If no charcoal, or an amount insufficient for a conventional radiocarbon determination was obtained, a second and then third quadrant was excavated in the same manner. No more than three quadrants were excavated from one individual feature. In several cases, sufficient charcoal was obtained from less than one whole quadrant, and excavation ceased.

Once excavation was completed, plans and stratigraphic sections were drawn and images taken. Additional notes about hearth structure and context were also made. Hearths were back-filled and reconstructed as close to their pre-excavation state as possible. A nail was deposited in each excavated section of all hearths to indicate prior excavation. Charcoal samples were stored in labeled aluminium foil packets and then placed in zip lock plastic bags.

Raw Material Source Identification and Mapping

To establish the lithic terrain (Elston 1990:155) of the study area, a survey for stone sources was undertaken. Potential sources of stone in the study area suitable for the production of artefacts were identified on the 1:250,000 Menindee geological map sheet (New South Wales Department of Mines 1969). An intensive pedestrian survey of these was undertaken to locate worked outcrops and sources of stone. While the scale of the geological maps is coarse and many smaller outcrops are not mapped at a fine scale, these provide an indication of the prospective location of potential lithic sources.

Once a worked outcrop was located, its position was recorded with the DGPS and images taken. Notes describing physical character, type and nature of raw material and evidence for stone working were made. To further characterise the range of raw material types available at each source, ten non-worked samples of stone were collected and flaked to assess internal structure and flaking properties. Less concentrated lithic sources, such as stream gravel and gibber pavements were also recorded and characterised according to the same methodology as that for outcrop sources.

Approaches to the Analysis of Stone Artefact Assemblage Composition

The features used to characterise stone artefact assemblage composition include assemblage size, raw material type representation, the degree of raw material reduction and utilisation, and the manufacture and maintenance of retouched tools. The following describes the use of these features.

Assemblage Size

While there is no ideal sample size for stone artefact assemblages, it is generally accepted that small assemblages are prone to the effects of outliers and have too few examples within each class of artefact for meaningful comparison. This may also result in different raw materials being combined for analysis thus compromising comparisons between raw material classes. These issues are especially pertinent to analyses of variance based upon measures of central tendency such as that employed in this study. A minimum assemblage size was therefore set at 3000 pieces. This figure was chosen because it was clear from reconnaissance that each of the four sampling locations contained at least this many artefacts above 20-millimetres in maximum dimension and that this was an achievable total given the time and labour constraints within which the project was conducted.

Quantifying Raw Material Proportion

The proportion of raw materials is calculated in three different ways. The standard method is to count each individual piece. However, for raw materials that produce large amounts of shatter during reduction, simply counting pieces irrespective of their technological attributes may over-estimate the importance of some materials in the overall raw material economy (Holdaway and Stern 2004). Two other methods can be used to confirm or refute the validity of basic raw material counts. The first is calculating the total volume (maximum length x maximum width x maximum thickness) of each raw material. However, as is the case with basic counts, volume does not consider the technological composition of assemblages. The second measure, the Minimum Number Flakes (MNF) calculation, is a technological measure of raw material proportion that considers the relative abundance of flakes.

Table 5.6 Description of artefact categories used for the initial quantification of the composition of the Pine Point-Langwell assemblages.

Category	Description
Broken Flake	Angular fragment, broken flake, complete split, distal flake, medial flake, proximal flake, proximal split
Broken Tool	Angular fragment tool, broken tool, complete split tool, distal tool, medial tool, proximal tool
Complete Bipolar Flake	Complete bipolar flake
Complete Flake	Complete flake
Complete Tool	Complete tool
Core	Core, bipolar core
Ground Stone	Milling slab, mortar, muller, pestle
Ground Stone Fragment	Milling slab fragment, muller fragment
Hammerstone	Hammerstone

Quantifying Technological Composition

There are many different ways to quantify the technological characteristics of lithic assemblages. Many of these are concerned with deriving estimates of the number of flakes and tools represented in assemblages (Hiscock 2001, Shott 2000). Detailed consideration of these is important because initial quantification guides further analysis.

The initial quantification of assemblage composition consists of basic counts of different technological categories. The categories utilised in this analysis are described in Table 5.6.

Measuring the Intensity of Raw Material Utilisation

The intensity with which blocks of stone were worked influences the overall composition of stone artefact assemblages. The intensity of raw material utilisation reflects the complex interplay between raw material distribution, form and availability, occupation duration and mobility. A number of ratios are employed in Chapter Eight to describe the intensity of raw material utilisation. These are based on the following categories:

MNF (Minimum Number of Flakes) – Count of all artefacts with a platform or point of force application. All complete flakes, complete tools, proximal flakes, proximal tools, half of complete split flakes, complete split tools and proximal splits

MNT (Minimum Number of Tools) – Count of all retouched items with a platform or point of force application. All complete tools, proximal tools, and half complete split tools.

Cores – All cores, excluding bipolar cores.

Summary

The archaeological techniques outlined in this chapter reflect the overall aim of the research to demonstrate the potential of surface deposits to yield information on long-term patterns in human behaviour. To do this requires:

- The establishment of occupational chronologies for surface deposits
- The mapping of artefact distributions over large spatial areas
- The recording a of minimum of several thousand stone artefacts per assemblage
- The application of techniques to quantify assemblage composition

The techniques applied during this study permitted the recording of high precision archaeological data comparable to the research conducted by WNSWAP in other areas of western New South Wales. These techniques are well established and focus on reducing errors and the effects of inter-observer bias through the computerisation of primary data recording.

The study area is situated in a different geomorphic context to the WNSWAP field areas. Unlike these, plains and low hills rather than the slopes and valleys of the rangelands dominate the Pine Point-Langwell study area. In many respects, the Pine Point-Langwell fieldwork is a test of the WNSWAP approach outside of the geomorphic context in which it was developed. This has necessitated minor modifications in the primary sampling of artefact distributions. These differences are illustrated by a comparison of the Pine Point-Langwell methodology with that outlined by Holdaway *et al.* (2000). A further difference is in the detail of geomorphic recording between the present study and WNSWAP. No direct dating of the sedimentary record of the Pine Creek and Rantyga Creek alluvial systems was undertaken due to the financial constraints involved with thermoluminescene and optically stimulated luminescence dating. The following chapter details the hearth excavation programme and establishes occupational chronologies for the Pine Point-Langwell sampling areas.

Chapter Six: Establishing a Chronological Context for the Pine Point-Langwell Surface Archaeological Record

Introduction

In Chapters One and Two it is argued that the chronological context of individual assemblages must be established prior to undertaking comparisons of assemblage composition. The chronological contexts of assemblages vary according to geomorphic context and occupation history. Research in the rangeland areas of western New South Wales demonstrates that assemblages derived from different land surfaces may not have contemporaneous chronological contexts (Holdaway *et al.* 2002a). Significant temporal variation in chronological context can be expected over spatial areas as small as several hundred metres, particularly in alluvial contexts that contain a range of terrace and floodplain land surfaces. This is the result of the complex geomorphic histories of landscapes in western New South Wales and the variable temporal character of occupation (Fanning and Holdaway 2001a, Holdaway *et al.* 2002a).

Aggregate deposits cannot be assumed to be contemporaneous or belong to the same behavioural system. In fact, the arid zone surface record is rarely, if ever, available to archaeologists at a resolution that permits the identification of fine grained contemporaneous assemblages. Despite this, some archaeologists continue to assume that the record documents a series of related assemblages and places representative of a behavioural system/s, often without the aid of absolute chronologies (e.g., Barton 2001). The temporal comparability of assemblages and locations is assumed rather than demonstrated.

Establishing occupational chronologies for surface deposits is one method for investigating the temporal nature of assemblage comparability. This will not establish the fine temporal resolution required for the identification of assemblages belonging to the same settlement system. Instead, it will provide coarse temporal units that permit the comparison of assemblages that have accumulated over similar or varied periods of time. The focus then becomes the investigation of long-term patterns in occupation and the spatial and temporal variation of assemblage structure and formation. This approach allows issues of diachronic variability in occupational chronologies, raw material utilisation and artefact discard within the overall context of place use histories to be addressed.

In this chapter, occupational chronologies are developed for the two alluvial systems from which the four assemblages investigated in this study are derived. The approach employed in the Pine Point-Langwell study closely follows that developed by WNSWAP (described in Chapter Two). As part of WNSWAP a detailed geomorphological and sedimentological dating programme, utilising absolute dating techniques (optically stimulated luminescence, thermoluminescene and radiocarbon dating) was undertaken to establish the temporal history of sedimentary sequences in alluvial valley fills (Fanning and Holdaway 2001a). This has proved crucial to establishing the maximum age of the lagged surfaces upon which deflated artefact distributions are recorded. In addition, dating the sedimentary sequence has also established the temporal context of the erosional processes that have removed earlier archaeological deposits. The absence of a dated sedimentary history is an acknowledged weakness of the Pine Point-Langwell study. However, general landscape history models for western New South Wales (e.g., Fanning 1999, Wasson 1979), plus the landform-based selection of hearths for dating, provide the basis for developing relative chronological contexts, and hence occupational histories for the surface distributions investigated in the Pine Point-Langwell study area.

Hearth Selection Criteria and Excavation Programme

As discussed in Chapter Four, landform context provides general clues to the temporal character of artefact accumulation. Older landforms, such as high alluvial terraces and those adjacent to palaeo-channels and oxbows are likely to represent accumulation over longer periods of time than surfaces on or near the floodplains of present channels. This information was used to select the hearths for excavation and dating in the Pine Point-Langwell study. The dating of multiple hearths on the landforms upon which artefact distributions were recorded allows the construction of occupational chronologies and the broad characterisation of the temporal units represented in artefact accumulation. While the temporal resolution of these contexts would be significantly improved with the absolute dating of sedimentary sequences, the temporal record from Pine Point-Langwell still represents one of the more detailed concentrations of dated surface hearths in the Australian arid zone.

Although assemblages were recorded from four locations along two different alluvial systems, radiocarbon determinations from each location within the separate alluvial system are combined for description and discussion. These systems are referred to as the Pine Creek alluvial system and the Rantyga Creek alluvial system.

Hearth Survey

The Pine Creek system encompasses the CN1 and CN3 sampling locations. A total of 87 heaths were recorded during a pedestrian survey of a five-kilometre transverse on the northeast side of Pine Creek. A hearth survey was not undertaken on the southwest side of Pine Creek because of the large width of the creek channel, the floodplain geomorphic context of the area and the fact that neither of the CN1 or CN3 sampling areas extended onto the southwest side. The Rantyga Creek system encompasses the KZ1 and KZ2 sampling locations. A total of 35 hearths were recorded during a pedestrian survey of a two-kilometre transverse on both sides of Rantyga Creek. Both sides of Rantyga Creek were surveyed because the KZ1 sampling area extended across the creek channel and landforms were similar.

Selection and Excavation Criteria

The recording procedure employed during the survey was described in the previous chapter. The excavation potential of the hearths was assessed on the basis of the degree of erosion. Hearths that were eroding from remnant sediment and retained a tight cluster of heat retainers were assessed to have a high excavation potential, particularly if surface charcoal was visible. Hearths that had fully eroded from remnant sediment with dispersed heat retainers were assessed as having a low excavation potential because it is unlikely that charcoal would be preserved.

Hearth condition varied considerably, with no obvious spatial pattern. Little is known about the rate at which hearths are destroyed following initial exposure, however this is likely to reflect a range of variables including micro-geomorphic context, environmental conditions (i.e. rainfall), and modern landuse (i.e. stocking rates on pastoral properties) (Holdaway *et al.* 2002b, 2005). The majority of hearths in both alluvial systems were classified as eroding (i.e. completely exposed) (Table 6.1). The proportion of completely exposed hearths is slightly greater for the Rantyga Creek system than the Pine Creek system. Partially exposed hearths were the second most common condition in both alluvial systems, with the proportion of these in the Rantyga Creek system approximately 10% greater than that for the Pine Creek system.

Table 6.1 Number and percentage (in brackets) of hearth condition in the Pine Creek and Rantyga Creek alluvial systems. The majority of hearths in both alluvial systems were classified as eroding.

Condition	Pine Creek	Rantyga Creek
Dispersed	24 (27.6)	5 (14.3)
Eroding	35 (40.2)	15 (42.9)
Exposed	28 (32.2)	15 (42.9)
Total	87 (100)	35 (100)

Table 6.2 Number and percentage (in brackets) of hearth excavation potential recorded in the Pine Creek and Rantyga Creek alluvial systems. The majority of hearths in both alluvial systems were classified as having a high excavation potential.

Condition	Pine Creek	Rantyga Creek
High	36 (41.4)	17 (48.6)
Medium	23 (26.4)	12 (34.3)
Low	5 (5.7)	3 (8.6)
None	23 (26.4)	3 (8.6)
	87 (100)	35 (100)

Dispersed hearths, those where the heat retainers have been extensively scattered, account for the lowest proportion, with the proportion approximately twice as large for Pine Creek than for Rantyga Creek.

Approximately 50% of hearths in the Rantyga Creek system and 40% in the Pine Creek system were classified as having a high excavation potential (Table 6.2). The proportion of hearths with medium and low excavation potentials is also greater in the Rantyga Creek system. However, the proportion of hearths with a low excavation potential is greatest in the Pine Creek system, reducing the overall suite of hearths available for excavation. The better preservation of hearths in the Rantyga Creek system may reflect the less extensive ground exposure in this land system compared to the CN land system. Other factors to do with the frequency and intensity of livestock grazing may also contribute.

Hearths were selected for excavation from the pool of hearths classified with a high excavation potential. Within this pool spatial proximity to the stone artefact sampling areas was the main criteria considered in deciding which hearths to excavate. Secondary considerations included geomorphic setting and the presence of visible charcoal. As permitted a total of 30 hearths were excavated. A total of 16 hearths yielded adequate samples of charcoal for conventional radiocarbon dating. Further sections of this chapter concentrate on the hearths that yielded charcoal samples.

Description of Excavated Hearths at Pine Creek that Returned a Charcoal Sample

Hearth CN 7 is eroding from the margin of a remnant sediment island onto the apron of a scald situated on a terrace 120 metres from the present channel of Pine Creek. The surface of the feature consists of a compact cluster of predominantly fire-cracked quartz hearthstones with fragments of charcoal visible, and was deemed to have a high excavation potential. Excavation involved the removal of the southeast quadrant to a maximum depth of 14 centimetres where the base of the fire pit was encountered. A moderate amount of bioturbation was noted in the upper five centimetres. Hearthstones were present throughout the matrix and were sitting directly on the base of the fire pit. The deposit primarily consisted of darkened soil and charcoal was plentiful throughout. A seven gram sample of charcoal was collected.

Hearth CN 23 is eroding from an area of remnant sediment on a terrace approximately 20 metres from the present channel of Pine Creek. The surface consists of a cluster of predominantly quartz hearthstones bound by sediment. The southwest and northwest quadrants were excavated, exposing a sandy fire darkened matrix with hearthstones throughout. The base of the fire pit was encountered at approximately 12 centimetres. Charcoal was not abundant, however a sufficient charcoal sample of 6.7 grams was collected.

Hearth CN 24 is eroding from an area of remnant sediment on a terrace 20 metres from the present channel of Pine Creek. The feature consists of a cluster of mainly quartz hearthstones bound by sediment. Excavation involved the removal of the southwest and southeast quadrants, exposing a sandy fire-darkened matrix with quartz hearthstones throughout. The fire pit was encountered at approximately 12 centimetres. Charcoal was abundant, and a large charcoal sample of 22.2 grams was collected.

Hearth CN 25 is located several metres from CN 24. Surface hearth stones are mainly of quartz. The southwest and southeast quadrants were excavated, exposing a sandy fire-darkened matrix with quartz hearthstones throughout. The base of the fire pit was encountered at approximately 15 centimetres. Charcoal was not abundant, but a charcoal sample of 3.6 grams was collected.

Hearth CN 32 is eroding from an area of remnant sediment on a terrace above a large oxbow located 100 metres from the present channel of Pine Creek. Hearths CN 32, 33, 34, 35 and 36 are clustered within several metres of each other. The surface of CN 32 consists of a compact cluster of predominantly fire-cracked quartz hearthstones bound with sediment. With charcoal fragments and a fire-darkened matrix visible on the surface, the feature was deemed to have a high excavation potential. Excavation proceeded to the fire pit and involved the removal of the southeast and southwest quadrants, exposing a sandy fire darkened matrix with hearthstones and abundant charcoal throughout. A charcoal sample of 34.3 grams was obtained.

Hearth CN 33 consists of a compact cluster of fire-cracked quartz hearthstones bound with sediment. Excavation proceeded to the fire pit at 12 centimetres and involved the removal of the southeast, southwest and northwest quadrants. The deposit contained a sandy fire-darkened matrix with few heat retainers and adequate charcoal for a 4.7 gram sample.

Hearth CN 34 consists of fire-cracked quartz hearthstones bound with charcoal blackened sediment. Excavation commenced in the southwest quadrant and immediately revealed a rich deposit of charcoal. To minimise unnecessary impact to the feature, excavation was limited to a 15 x 15 centimetre area. Excavation ceased at 10 centimetres. The deposit contained a sandy fire-darkened matrix with quartz heat retainers. A 2.8 gram charcoal sample was obtained.

Hearth CN 35 consists of a compact cluster of fire-cracked quartz hearthstones bound with charcoal blackened sediment. Excavation commenced in the southwest quadrant and immediately revealed a rich deposit of charcoal. To minimise unnecessary impact to the feature, excavation was limited to a 15 x 15 centimetre area. Excavation ceased at a depth of 12 centimetres. The deposit contained a sandy fire-darkened matrix with quartz heat retainers. An adequate charcoal sample of 4.1 gram was obtained.

Hearth CN 36 is made up of fire-cracked quartz hearthstones bound with sediment. Excavation involved the removal of the majority of the feature, revealing a sandy matrix with occasional quartz heat retainers and sparse charcoal. The fire pit was encountered at seven centimetres and a charcoal sample of 2.2 grams was obtained.

Hearth CN 42 is eroding from the margin of a remnant sediment island onto the apron of a scald situated on the terrace close to the margin of the present floodplain of Pine Creek. The hearth consists of a compact cluster of predominantly fire-cracked quartz hearthstones. Large areas of charcoal and charcoal stained soil were visible on the surface and the feature was deemed to have a high excavation potential. Excavation involved the removal of a 20 x 20 centimetre area in the southwest quadrant. Abundant charcoal was immediately encountered in a sandy charcoal darkened matrix and a sufficient sample was quickly obtained. Hearthstones were present throughout the matrix. To avoid unnecessary disturbance to the feature, excavation ceased at a depth of 10 centimetres without reaching the fire pit. The feature appeared structurally similar to CN H 7, 23, 32 and it was thought that further damage did not warrant the limited knowledge that would be gained about the hearth structure. A charcoal sample of five grams was obtained.

Hearth CN 55 is eroding from remnant sediment on the edge of the terrace and the present floodplain of Pine Creek. The surface of the feature consists of fire-cracked quartz and clay heat retainers. Although no surface charcoal was visible, the tight clustering of hearthstones suggested the presence of charcoal in the deposit, therefore it was deemed to have a high excavation potential. Excavation involved the removal of the southeast quadrant to the fire pit, which was reached at a depth of 10 centimetres. The matrix of the deposit was rich in charcoal and contained few hearthstones and little structure of note. A large charcoal sample of 18.4 grams was obtained.

Hearth CN 56 is exposed on the scalded terrace surface. The feature consists of a large number of dispersed predominantly quartz hearthstones, with a smaller amount of quartzite fire cracked heat retainers and surface charcoal. Excavation commenced in the northeast quadrant, revealing an extremely rich charcoal deposit in a sandy matrix. A 7.1 gram charcoal sample was obtained from a 10 x 10 centimetre area.

Description of Excavated Hearths at Rantyga Creek that Returned a Charcoal Sample

Hearth KZ1 2 is exposed on the edge of a small eroding terrace 20 metres from the present channel of Rantyga Creek. The surface of the feature consists of several fire cracked quartz heat retainers bound with sediment. Small areas of charcoal stained soil were visible. Excavation involved the removal of the southwest quadrant. Abundant charcoal was discovered immediately beneath the hearth-stones. The deposit consisted of a charcoal stained sandy matrix. A 4.1 gram charcoal sample was obtained.

Hearth KZ1 3 is eroding from the margin of remnant sediment onto the apron of a scald approximately 50 metres from the present channel of Rantyga Creek. The hearth consists of a small number of quartz and clay hearthstones. Two minor areas of charcoal staining were visible. The northeast, southeast and southwest quadrants were removed to the fire pit at a depth of eight centimetres. The hearth contained little discernible structure, however clay heat retainers were encountered throughout the coarse sandy deposit. Charcoal was not abundant and was primarily obtained from beneath a clay heat retainer on the north edge of the southeast quadrant. However, a charcoal sample of four grams was obtained.

Hearth KZ1 7 is eroding from the margin of remnant sediment onto the apron of a scald 50 metres from the present channel of Rantyga Creek. The surface of the feature consists of a small number of clay heat retainers bound with sediment. Excavation involved the removal of the northeast and northwest quadrants, proceeding to the fire pit at a depth of eight centimetres. The hearth contained little structure, but clay heat retainers were encountered throughout the coarse sandy deposit. A 7.7 gram charcoal sample was obtained.

Hearth KZ2 25 is eroding from the western hill slope over looking KZ2 and therefore has a different geomorphic setting to the other hearths excavated along Rantyga Creek. The surface of the feature consists of a cluster of fire-cracked quartz heat retainers with exposed surface charcoal. Excavation removed the southwest quadrant to a depth of 10 centimetres. The deposit consisted of numerous stratified quartz heat retainers and abundant charcoal, with a charcoal sample of 6.7 grams obtained.

Summary - Hearth Structure

Hearths in the Pine Point-Langwell study area generally have shallow deposits. These consist of a sandy matrix with occasional, usually quartz, heat retainers and small amounts of charcoal. There is no structural evidence that suggests the hearths were reused over long periods of time. Heat retainers were typically separated from the base of the fire pit by several centimetres of soil, but were also often resting directly upon the fire pit. Charcoal was usually found directly underneath the heat retainers and very rarely as distinct seams. Multiple layers of charcoal were not recorded in any of the hearths.

In addition to hearth structure, the geomorphic history of the landscape means that it is unlikely that hearths were reused for long periods of time following their initial construction. Hearths are not prominent features in the landscapes of western New South Wales. They are primarily visible due to the large amount of erosion that has occurred during the pastoral period. In any case the remains of hearths visible in the present represent the bottom of the structure, and this would not be visible without the work of erosion. Therefore, both the internal structure of the hearths and geomorphic processes mean that it is unlikely that hearths were reused or in use for long periods of time during the past.

Radiocarbon Determinations from the Pine Point-Langwell Hearths

Sufficient wood charcoal for conventional radiocarbon determinations was obtained from all 16 hearths. All charcoal samples were submitted to the University of Waikato Radiocarbon Dating Laboratory in New Zealand.

Sample Preparation

Samples submitted for dating were first cleaned by wet sieving to remove soil particles and loose material, and then by air drying the material retained in the sieve. Charcoal was then picked out from roots, seeds and other non-datable material using tweezers, and sent to the University of Waikato Radiocarbon laboratory. Here, the samples were washed in hot 10% HCL, rinsed with deionised water, and treated with hot 1% NaOH. The NaOH insoluble fraction was then treated in with hot 10% HCL, filtered, rinsed and dried. This process removes potential contaminates from the samples.

Results

The 16 radiocarbon determinations from the Pine Point-Langwell heat retainer hearths are presented in Table 6.3. The determinations are sorted in chronological order from youngest to oldest in both alluvial systems. Calibrated determinations at the 95% probability level are presented in Table 6.4.

Table 6.3 Radiocarbon determinations on charcoal from heat retainer hearths on Pine Point and Langwell Stations. Weight refers to cleaned charcoal.

Hearth ID	Alluvial System	Lab ID	DC13	% Modern	Result (year BP)	Wt (g)
CN H 42	Pine Ck	Wk-9994	-23.8 ± 0.2	96.8 ± 0.6	261 ± 49	5
CN H 34	Pine Ck	Wk-12322	-24.1 ± 0.2	94.5 ± 0.7	458 ± 58	2.8
CN H 35	Pine Ck	Wk-12323	-23.5 ± 0.2	94.2 ± 0.6	481 ± 55	4.1
CN H 36	Pine Ck	Wk-12324	-23.0 ± 0.2	93.8 ± 0.6	516 ± 52	2.2
KZ2 H 25	Rantyga Ck	Wk-12326	-22.2 ± 0.2	93.0 ± 0.5	584 ± 46	6.7
CN H 56	Pine Ck	Wk-12325	-24.6 ± 0.2	90.9 ± 0.5	771 ± 46	7.1
KZ1 H 2	Rantyga Ck	Wk-9995	-23.4 ± 0.2	90.0 ± 0.8	848 ± 69	4.1
CN H 55	Pine Ck	Wk-10282	-23.9 ± 0.2	89.6 ± 0.5	886 ± 47	18.4
CN H 23	Pine Ck	Wk-10280	-23.0 ± 0.2	89.3 ± 0.6	910 ± 52	6.7
KZ1 H 7	Rantyga Ck	Wk-10832	-22.1 ± 0.2	88.7 ± 0.6	959 ± 51	7.7
CN H 25	Pine Ck	Wk-12320	-21.9 ± 0.2	88.7 ± 0.7	967 ± 62	3.6
CN H 24	Pine Ck	Wk-12319	-24.5 ± 0.2	88.3 ± 0.5	1002 ± 48	22.2
CN H 33	Pine Ck	Wk-12321	-23.9 ± 0.2	85.6 ± 0.6	1247 ± 56	4.7
CN H 32	Pine Ck	Wk-10281	-23.2 ± 0.2	81.4 ± 0.5	1653 ± 52	34.3
CN H 7	Pine Ck	Wk-9993	-23.5 ± 0.2	80.5 ± 0.8	1747 ± 76	7
KZ1 H 3	Rantyga Ck	Wk-10831	-23.2 ± 0.2	77.9 ± 0.7	2004 ± 73	4

Table 6.4 Calibrated determinations at the 95% probability level for the Pine Point-Langwell study area.

Hearth ID	Alluvial System	Lab ID	Result (year BP)	95.4% Probability Level
CN H 42	Pine Ck	Wk-9994	261 ± 49	1480AD (75%) 1690AD 1730AD (17.1%) 1810AD 1930AD (3.3%) 1950AD
CN H 34	Pine Ck	Wk-12322	458 ± 58	1320AD (2.8%) 1350AD 1390AD (79.9) 1530AD 1550AD (12.6%) 1640AD
CN H 35	Pine Ck	Wk-12323	481 ± 55	1300AD (9.2%) 1360AD 1380AD (83.2%) 1520AD 1590AD (3%) 1630AD
CN H 36	Pine Ck	Wk-12324	516 ± 52	1300AD (26.1%) 1370AD 1380AD (69.3%) 1480AD
KZ2 H 25	Rantyga Ck	Wk-12326	584 ± 46	1290AD (95.4%) 1430AD
CN H 56	Pine Ck	Wk-12325	771 ± 46	1160AD (95.4%) 1300AD
KZ1 H 2	Rantyga Ck	Wk-9995	848 ± 69	1030AD (95.4%) 1290AD
CN H 55	Pine Ck	Wk-10282	886 ± 47	1020AD (95.4%) 1250AD
CN H 23	Pine Ck	Wk-10280	910 ± 52	1020AD (95.4%) 1250AD
KZ1 H 7	Rantyga Ck	Wk-10832	959 ± 51	990AD (95.4%) 1210AD
CN H 25	Pine Ck	Wk-12320	967 ± 62	970AD (95.4%) 1220AD
CN H 24	Pine Ck	Wk-12319	1002 ± 48	900AD (2.7%) 920AD 960AD (92.7%) 1170AD
CN H 33	Pine Ck	Wk-12321	1247 ± 56	660AD (95.4%) 900AD
CN H 32	Pine Ck	Wk-10281	1653 ± 52	250AD (95.4%) 540AD
CN H 7	Pine Ck	Wk-9993	1747 ± 76	80AD (95.4%) 440AD
KZ1 H 3	Rantyga Ck	Wk-10831	2004 ± 73	210BC (95.4%) 140AD

Scale of Temporal Analysis - Occupational chronologies for the Two Alluvial Systems

The 16 radiocarbon determinations date a series of events. Considered in isolation, each determination only indicates that people visited the study area at a particular time during the past. To build an occupational chronology, patterns must be identified among multiple determinations. A Bayesian analysis of the determinations indicates that there is strong evidence for the existence of multiple episodes of hearth construction separated by hiatuses when tested against the null hypothesis of a single occupation phase.

A calibrated probability distribution plot for determinations from both of the alluvial systems is presented in Figure 6.1. It is clear that the radiocarbon determinations do not form a continuous sequence, but cluster into five main groups. The oldest is represented by a single determination from KZ1 H3 at approximately 2,000 years cal BP. The second group consists of two hearths with determinations between 1,500 to 1,700 years cal BP. The third group consists of one hearth with a calibrated determination between 1,050 to 1,300 years cal BP. Seven determinations form the fourth group, with determinations spanning 750 to 950 years cal BP. The fifth group consists of five determinations that span 550 to 350 years cal BP.

Bayersian analysis can be used to test the null hypothesis that the determinations represent a single phase of hearth construction against the alternative that the determinations represent multiple phases of occupation. Using Datelab v. 1.2 software (M. Jones and Nicholls 2003), the mean likelihood statistic can be calculated for H_0 and H_1. The two statistics are then divided (H_0/H_1) and the result (the Bayes Factor) compared to values published by Raftery (1996) and presented here in Table 6.5. Values greater than 12 demonstrate strong evidence for multiple groups of determinations. The Bayes value is highest for (A:B) a single versus multiple (five) phase hearth construction (Tables 6.6 - 6.7), indicating that there is very strong evidence for multiple (five) phases of hearth construction rather than the null hypothesis that the hearths represent a single phase of hearth construction.

There are a number of possible explanations why the hearths should fall into multiple phases. The most obvious is that hearths dating to periods between the phases have been removed by erosion. This argument can be discounted if the spatial distribution of the hearths is considered (Figure 6.2). As the CN H 32 – CN H 36 cluster of hearths demonstrates, this pattern does not represent the selective removal through erosion of hearths of certain ages. CN H 32, 33, 34, 35 and 36 are located within a ten metre radius of each other on the surface of the major Pine Creek terrace. Determinations from these hearths fall within phases one, five, and seven of hearth construction (Table 6.6). Therefore, it may be concluded that for much of the last 1600 years the surface along the northern terrace of Pine Creek has remained relatively stable. This suggests that the discontinuous hearth record, at least on the terrace, is not the product of erosion.

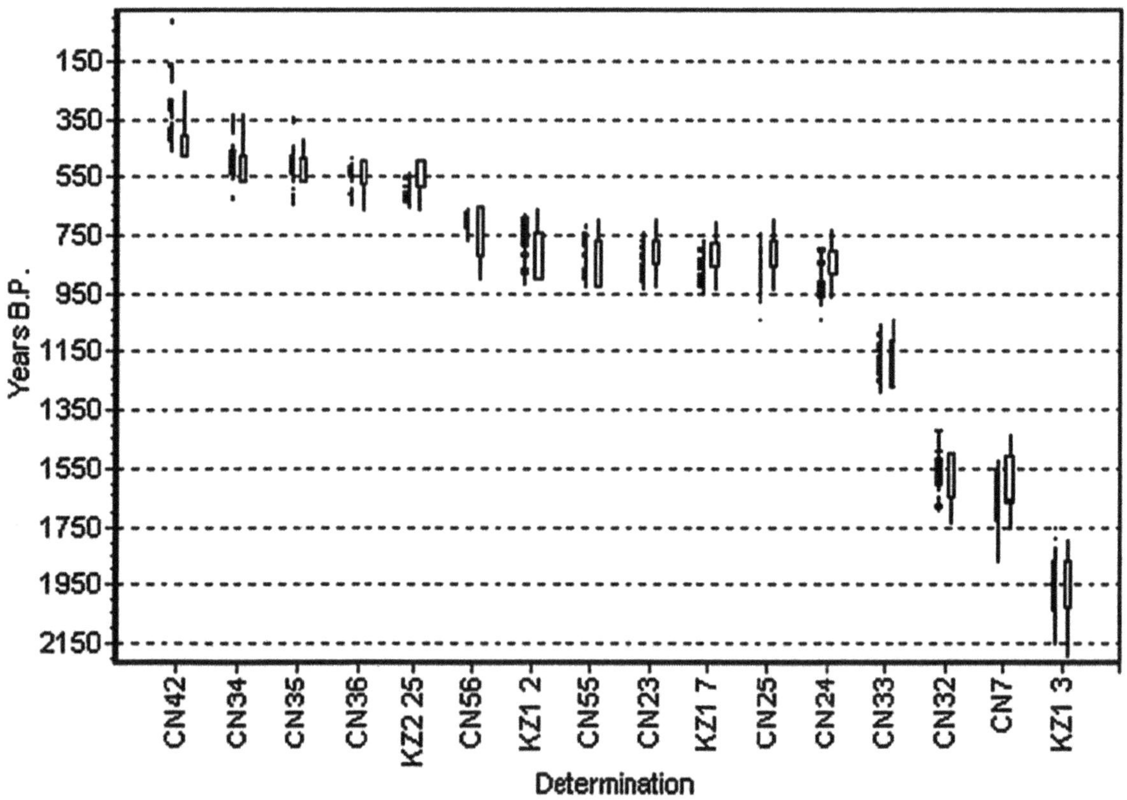

Figure 6.1 Calibrated radiocarbon determination plot for the Pine Point-Langwell hearths. Likelihood is represented by the solid black bars and the posterior by white fill.

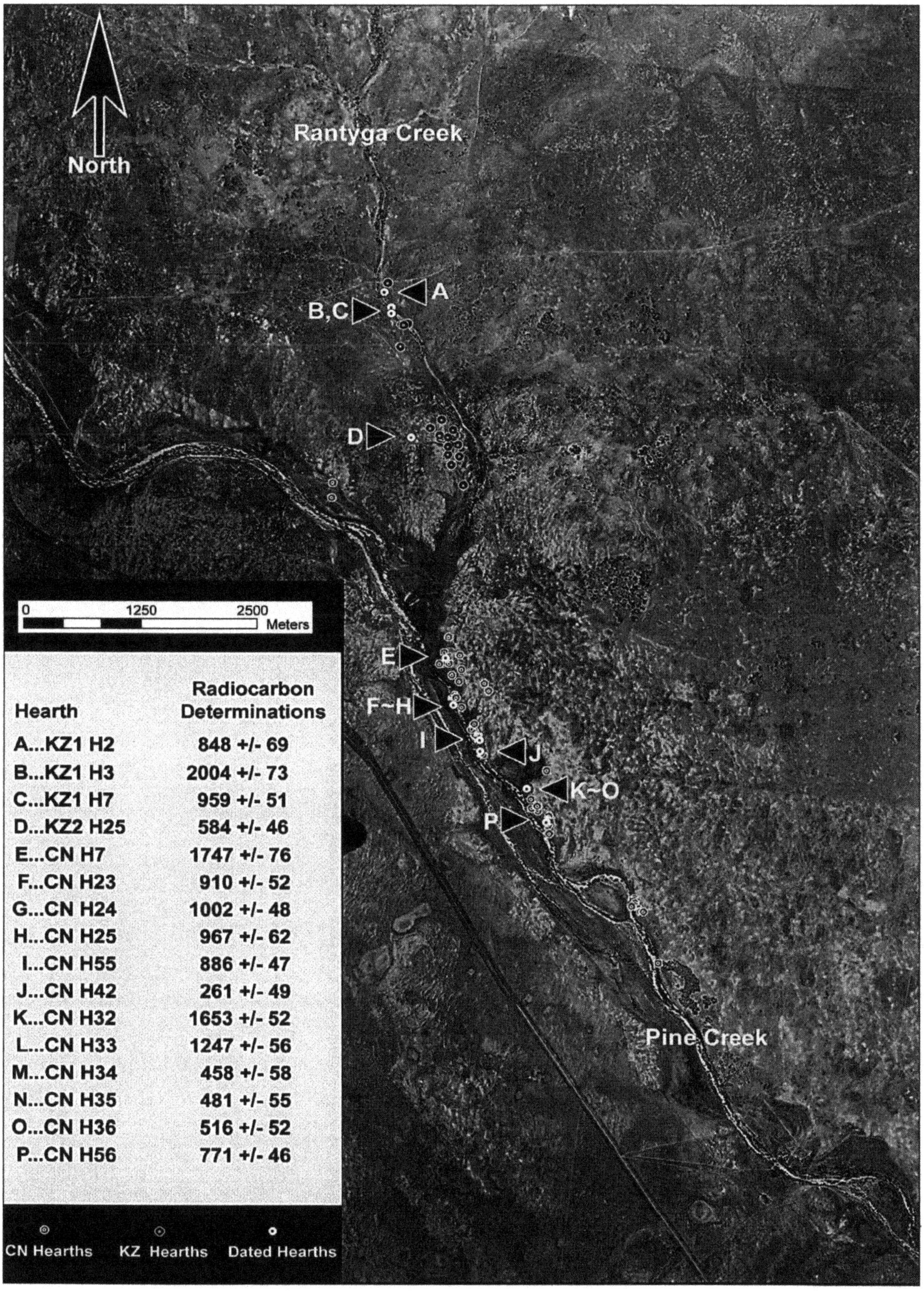

Figure 6.2 Distribution of dated hearths from the Pine Creek and Rantyga Creek alluvial systems.

Table 6.5 Interpretation of Bayes factor Model 2 v. Model 1, from Rafferty (1996).

Bayes factor	Support for (Model 1 vs. Model 2)
< 1	Supports Model 1
1 to 3	Barely worth mentioning
3 to 12	Positive
12 to 150	Strong
>150	Very strong

Table 6.6 Single versus multiple phases of hearth construction tested for the Pine Point-Langwell determinations. Phase A represents the null hypothesis of a single phase, Phase B (five), Phase C (three) and Phase D (two).

Date	Result	Error	Phase A	Phase B	Phase C	Phase D
CN42	261	49	1	1	1	1
CN34	458	58	1	1	1	1
CN35	481	55	1	1	1	1
CN36	516	52	1	1	1	1
KZ2 25	584	46	1	1	1	1
CN56	771	46	1	3	3	1
KZ1 2	848	69	1	3	3	1
CN55	886	47	1	3	3	1
CN23	910	52	1	3	3	1
KZ1 7	959	51	1	3	3	1
CN25	967	62	1	3	3	1
CN24	1002	48	1	3	3	1
CN33	1247	56	1	5	5	3
CN32	1653	52	1	7	5	3
CN7	1747	76	1	7	5	3
KZ1 3	2004	73	1	9	5	3

Table 6.7 Mean likelihood ratio for Single versus multiple phases of hearth construction tested for the Pine Point-Langwell determinations.

H_0 vs. H_1	Mean Likelihood Ratio Value
A:B	631
A:C	10.8
A:D	0.253

It is difficult to identify the reasons behind the repeated cessation of hearth construction in the sampling areas along Pine and Rantyga Creeks. The hiatuses may indicate that human-environment relationships changed on several occasions within the last 2000 years in the study area. These changes may represent both social and physical factors. At present palaeoenvironmental records for western New South Wales are too coarse in resolution to offer a clear explanation (see Holdaway *et al.* 2002a). While Rantyga Creek demonstrates an older occupation history than the section of Pine Creek sampled in this study, the record is also one of discontinuous occupation. Like the Pine Creek occupational chronology, the majority of the determinations from the Rantyga Creek alluvial system (three of four) are more recent than 1000 years BP. In this respect the two occupation chronologies are broadly similar.

Cupper (1998) identified major changes in vegetation structure in the pollen record of lakes from southwest New South Wales during the last several thousand years. He argued that these changes reflected increasing aridity or greater seasonal variation in the climate during the late Holocene. Therefore, while the available data is not definitive, it does suggest a degree of environmental and climatic variability in western New South Wales during the period represented by the Pine Point-Langwell occupational chronology. Following an extensive review of the palaeoenvironmental literature, Holdaway *et al.* (2002a) suggested a possible correlation between the occupational hiatus at Stud Creek in far northwest New South Wales and the evidence for global climatic change. The hiatus at Stud Creek coincides with a period of climatic variability between 1150 and 800 years BP known as the Medieval Warm Period, where colder temperatures interrupted brief warm periods. Interestingly, none of the Pine Point-Langwell determinations fall to within the period of 1150 to 950 years BP. This may reflect a local response to this period of environmental and climatic oscillation.

Although the palaeoenvironmental record is sparse for arid Australia, analysis of the available data requires a regional perspective. The temporal signature at Stud Creek is only one occupational chronology. While there is a correlation between the archaeological chronology and the palaeoenvironmental record at Stud Creek, different patterns may be identified in other areas of western New South Wales. It is only through the dating of large numbers of hearths from across the region that local responses to climatic oscillation will be evident.

Determinations and Landforms – Spatial Patterns

One of the main aims of the dating programme is to establish occupational chronologies across different landforms in the sampling areas. This also provides an indication of the periods during which artefact discard may have taken place. It is difficult to establish the maximum age of the surfaces upon which artefacts are distributed without absolute dates from the sedimentary record. However, radiocarbon determinations from the hearths combined with general geomorphic models established for similar areas of western

New South Wales (e.g., Fanning 1999, 2002, Wasson 1979) can provide hints as to the relative chronological contexts of archaeological deposits in the alluvial systems. If sufficient radiocarbon determinations are obtained from individual landforms it is possible to develop occupational histories for different landforms. The temporal record from Pine Point-Langwell allows broad scale spatial comparisons between locations with established occupational chronologies.

All hearths within the vicinity of the CN1 and CN3 sampling areas are situated on the top of the creek terrace. Extensive investigation of the floodplain and terrace slope failed to locate any hearths in these locations. This fact reflects the depositional nature of these surfaces obscuring archaeological distributions with sediment and vegetation. The terrace surface above the floodplain represents a relatively long record of discontinuous occupation, as demonstrated by the close proximity of hearths CN H 32 (1653±52 years BP), CN H 33 (1247±56 years BP), CN H 34 (458±58 years BP), CN H 35 (481±55 years BP) and CN H 36 (516±52 years BP) (Figure 6.2). This demonstrates that hearths of different ages can be found on the same surface.

Another factor to consider when assessing spatial and temporal patterns in the occupational chronology is the exposure of individual hearths. Fifteen of the hearths were recorded as eroding from remnant sediment, and one as being at an advanced state of heat retainer dispersal on completely exposed hard-pan scald. Interestingly the fully exposed hearth, CN H56 (Wk-12325) returned a determination of 771±46 years BP adding further weight to the argument that hearth age is not correlated with the extent of erosion. In any case, evidence from other areas of western New South Wales suggests that the present condition of landforms along alluvial systems largely reflect accelerated erosion during the last 150 years (Fanning 1999). Therefore, the hearth age and degree of preservation are not necessarily correlated in the study area.

Although far fewer determinations were obtained from the Rantyga Creek alluvial system, similar relationships between hearth exposure and age can be suggested. All four hearths were recorded as eroding from remnant sediment islands rather than being completely exposed. Three of the four hearths were recorded on the distal floodplain of Rantyga Creek, and there is no evidence of geomorphic processes that have selectively removed hearths of specific ages. Indeed, the oldest hearth in the study area KZ1 H3 at 2004 ± 73 years BP (WK 10831) is located on the distal floodplain of the Rantyga Creek alluvial system. This hearth is located within 100 metres of other hearths on the distal floodplain that returned determinations of 959 ± 51 and 848 ± 69 years BP. The fourth hearth KZ2 H25 (WK 12326) is situated on a gentle slope near the edge of the distal floodplain and returned a determination of 584 ± 46 years BP.

Discussion – Regional Chronologies

The Pine Point-Langwell radiocarbon determinations point to some interesting temporal patterns. There is now a large database of radiocarbon determinations from heat retainer hearths in other parts of western New South Wales (Holdaway *et al.* 2002a, 2005). The uniqueness of the Pine Point-Langwell patterns can be compared to occupational chronologies from other areas.

Regional Radiocarbon Determinations

As discussed in earlier chapters, selected parts of western New South Wales around the Willandra Lakes World Heritage Area and the Menindee Lakes have been the focus of a considerable amount of archaeological work. Radiocarbon determinations on a range of archaeological material in conjunction with the dating of geomorphic features indicate human occupation extending beyond 40,000 years BP (Bowler and Price 1998, Johnston and Clark 1998). Therefore, in the overall context of the wider western New South Wales region, the Pine Point-Langwell determinations are relatively recent in age. They are consistent with other large samples of hearth determinations from Stud Creek and Fowlers Gap (Holdaway *et al.* 2002a).

Hearth Determinations

The Western New South Wales Archaeological Programme (WNSWAP) has a developed a large database of radiocarbon determinations from four research areas to the north and east of Pine Point-Langwell. The largest suite of hearth determinations comes from the Fowlers Gap Arid Zone Research Station in the northern Barrier Range, approximately 160 kilometres north of the Pine Point-Langwell study area. A total of 45 determinations were obtained from six different sampling areas across the 38,850 hectare property.

When the Fowlers Gap determinations are combined they indicate a number of interesting temporal patterns across the study area (Figure 6.3). As with the Pine Point-Langwell occupational chronology there is a group of determinations before 500 years BP and then another group between 750-950 years BP. Unlike Pine Point-Langwell there is no gap in hearth construction between 1150-1350 years BP at Fowlers Gap. However, there are only a small number of hearths representing this period, thus suggesting a period of low intensity hearth construction.

Another major difference between the two sets of determinations is the age of the oldest hearths. The oldest hearth at Fowlers Gap belongs to the Fowlers Creek sampling area, and returned an age estimate of approximately 6,000 years BP. Additionally, a further eight hearths in this same area provided determinations between 3,000 – 4,000 years BP. This is very different situation to Pine Point-Langwell, where 2,000 years BP is the oldest age estimate from KZ1 H3 (WK 10831) in the Rantyga Creek alluvial system. The old determinations from the Fowlers Creek sampling area reflect the

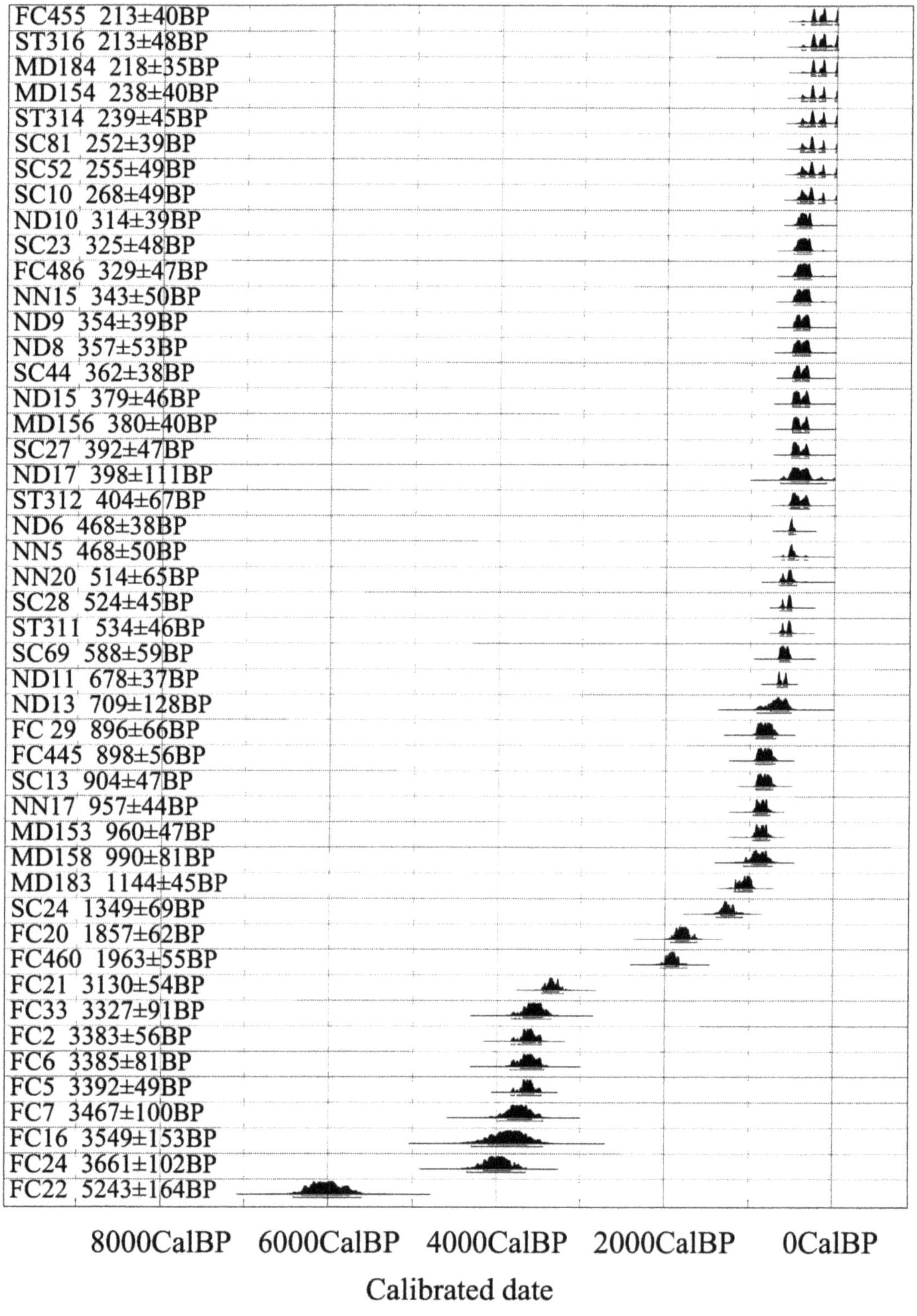

Figure 6.3 Calibrated radiocarbon determinations from hearths on Fowlers Gap Station. The column on the left contains the hearth code and the un-calibrated determination (Information provided by Dr. Simon Holdaway and Dr. Patricia Fanning).

geomorphological circumstances of hearth preservation, with erosion removing earlier hearths in the other sampling areas at Fowlers Gap (Simon Holdaway pers. comm).

It is very interesting that some similar temporal patterns in hearth construction are evident at Pine Point-Langwell and Fowlers Gap. As discussed earlier, it is difficult to identify the reasons behind this pattern. Current palaeoenvironmental data is too coarse for the identification of clear environmental signatures. The significance of the pattern for the present study lies in the confirmation that the pattern of significant hiatuses in hearth construction, first identified by Holdaway (*et al.* 2002a) at Stud Creek, is repeated in other areas of western New South Wales. Therefore, the records of both Pine Point-Langwell and Fowlers Gap are of discontinuous occupation punctuated by periods when hearth construction ceased or was very much reduced. This has significant implications for the way that the surface archaeological record is interpreted. Firstly, the record has a dynamic history, with the present distribution reflecting both the locations of human behaviour in the past and the operation of geomorphic processes (see Holdaway *et al.* 2002a). Some places preserve an old record, but geomorphic processes may result in a more recent record in others. Secondly, these factors mean that contemporaniety between different areas must be demonstrated rather than assumed. For example hearths separated by several hundred metres may be many of hundreds and even thousands of years different in age. Thirdly, the record cannot be interpreted as representing synchronic lifeways unless the temporal dimension of the occupational histories of different places is ignored. Occupational histories indicate that not all places across a landscape were occupied at the same time.

Summary

The surface record represents the accumulation of material from multiple events. The dynamic occupational chronologies identified at several locations across western New South Wales and now also at Pine Point-Langwell demonstrate that this record of multiple events is punctuated by hiatuses with no evidence of hearth construction. This is interpreted as repeated abandonment and suggests that there is little continuity in behavioural adaptation (certainly in terms of place use) over the duration of the various occupational chronologies now established for western New South Wales. Taken as a time-averaged record, the material on the deflated Pine Point-Langwell deposits cannot be interpreted as representing a system or model. Instead, it is argued that it is more appropriate to investigate historical variability in the accumulation of materials at different locations, rather than viewing these locations as parts of the same system. The occupational chronologies have demonstrated temporal variability. The challenge is to now investigate stone artefact assemblage accumulation without recourse to synchronic interpretations, which impose uniform sets of behaviours, because clearly such simple explanations ignore the complexity of the record.

Chapter Seven: Raw Material Representation and Technological Description

Introduction

The archaeological record of Pine Point-Langwell can be investigated at a number of spatial and temporal scales. Within these scales it is possible to discuss both continuity and discontinuity in the character of occupation. Over the long-term the occupational chronology for the Pine Point-Langwell study area is one of discontinuity with periods of hearth construction separated by significant hiatuses. At the same time it is possible that other factors have resulted in behavioural continuity over the long-term. These factors may function as constraints on the level of behavioural variability, resulting in the formation of clear accumulated patterns. One example of such a factor is the distribution and availability of lithic raw materials in the study area resulting in the formation of a clear pattern in raw material reduction over long periods of time.

The geological context of potential sources of raw material in the Pine Point-Langwell study area suggests that these have remained constant throughout the period represented in the human occupational chronology. Both silcrete and quartz occur on residual surfaces that are not exposed to the same geomorphic processes as those in the creek valley. While there is evidence that a small amount of raw material was transported into the study area, much of the raw material represented in the assemblages is available at local sources. Beyond these constraints, variability will come from economic factors such as distance decay relationships between raw material sources and areas of off-source reduction. If a distance decay relationship is observed, this may be an example of long-term factor conditioning behavioural continuity.

In this chapter, the raw material composition of the Pine Point-Langwell assemblages is described. The description of the assemblages begins with a summary of the potential sources of stone suitable for the production of artefacts available within the study area. The relative proportion of different raw materials in the assemblages is then quantified according to the methods outlined in Chapter Five. The representation of raw material is related to the range, form and distribution of potential sources of stone suitable for the production of artefacts. The chapter concludes with a consideration of the possible constraints that raw material availability, form and quality played in controlling variability between the assemblages. Once the effects of raw material are understood and controlled, it is possible to pursue an investigation of historical variability that may reflect the influence of other factors representative of the character of occupation.

The Occurrence and Form of Potential Raw Material Sources within the Pine Point-Langwell Study Area

The type, distribution and form of raw materials suitable for stone artefact manufacture are important factors that contribute to assemblage composition (Andrefsky 1994, Hiscock 1985). As reviewed in Chapter Three, these factors are at the forefront of many Australian stone artefact studies and are often emphasised at the expense of other factors that influence assemblage composition (e.g., occupation duration). To get beyond the effect of raw material availability alone still requires that the range and distribution of potential raw material sources be quantified. With this in mind, the character of potential raw material sources for the stone artefacts in the Pine Point-Langwell assemblages is described.

Three types of raw material suitable for stone artefact manufacture occur naturally within the study area: quartz, quartzite and silcrete. Figure 7.1 shows the distribution of potential raw material sources within the Pine Point-Langwell study area relative to the CN1, CN3, KZ1 and KZ2 localities.

Quartz

Quartz is a hard dense rock. It has formed in the Barrier Range as a result of the deposition of silica as quartz in the fractures of other rocks during metamorphism (Plimer 1994:48). Extensive quartz gibber pavements occur widely within the study area. However, unlike the Barrier Range to the north, where milky quartz vein outcrops are common (Plimer 1994), no worked vein quartz outcrops were found in the study area. Quartz gibber pavements are concentrated in the Enmore Hills six kilometres to the northwest of the Rantyga-Pine Creek confluence and on the foot slopes of the Sampson's Paddock Hills bordering Rantyga Creek three kilometres to the north. Quartz nodules also occur less commonly in gravel bars along sections of Pine Creek and as gravel layers in the sedimentary sequence of the Pine Creek Valley. The availability of these to the Aboriginal people who inhabited the area as sources for artefact production is unknown.

Quartz gibber nodules are mainly milky (white) to opaque (partially transparent) in appearance and a significantly smaller amount of crystal quartz nodules also occur. A smooth rounded cortex covers the exterior of the nodules. Test reduction of selected nodules indicated that the knapping quality varies considerably. Internal flaws in the structure of the nodules are common, causing them to

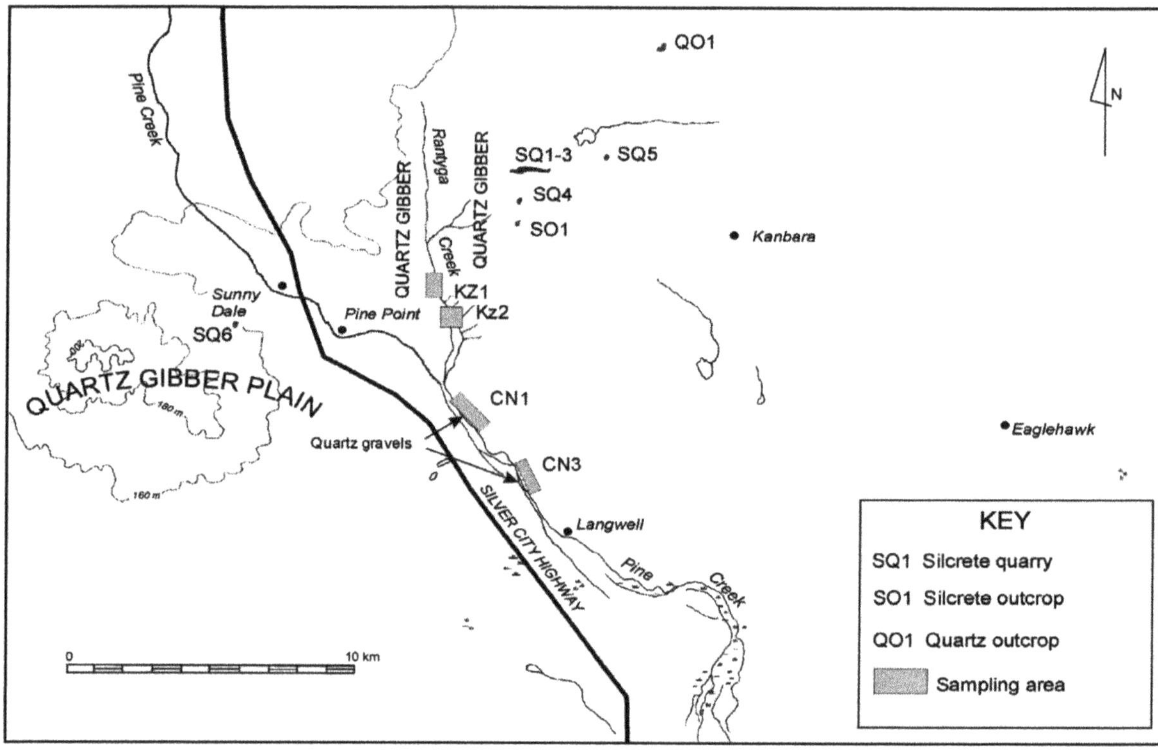

Figure 7.1 Distribution of potential raw material sources relative to the four sampling areas.

shatter into several or more angular blocks upon impact with a blow from a hammerstone. Dickson (1977:102) argues that translucence is the most important visual indication of the flaking qualities of quartz. From this it is inferred that crystal quartz and opaque quartz is of a higher quality than milky quartz in the study area. However, excluding crystal quartz that is very distinctive, it was difficult to separate opaque and milky quartz based upon transparency. In addition, the testing of nodules indicated that individual nodules were often both opaque and milky in parts. Due to these factors, opaque quartz and milky quartz are combined for analytical purposes. Conchoidal flakes, exhibiting a platform, bulb of percussion and termination can be produced through the freehand reduction of nodules free of internal flaws. No quantitative size data is available, however quartz nodules occur in a range of sizes, with the largest typically fist sized (approximately 10-15 centimetres in diameter).

Reef quartz outcrops are extremely rare, with only two isolated areas noted. One of these is situated on the top of a hill approximately six kilometres north of KZ1 while the other is a very minor outcrop located in the Enmore Hills on Sunnydale Station. Both outcrops exhibit large fractured blocks typically larger than 10 x 10 centimetres, with numerous internal flaws. This particular type of quartz may have a lower reduction utility compared to quartz gibber and it is expected to comprise a relatively small amount of the overall proportion of worked quartz. Artefacts were not noted at either of the vein quartz outcrops.

Quartzite

Quartzite is a hard dense rock formed by the recrystallisation through metamorphism of quartz grains in sandstone (Cochrane *et al.* 1991:18). Quartzite is the least ubiquitous of the raw materials found within the Pine Point-Langwell study area. It occurs as isolated nodules within the larger quartz gibber pavements. Like quartz gibber, a smooth rounded cortex covers the external surface of quartzite nodules. Some quartzite can be similar in appearance to coarse-grained silcrete and the two have sometimes been confused (Bird 1985). Quartzite in the study area is characterised by a coarse quartz grained matrix that is sugary in appearance and easily distinguishable from clast and non-clast silcrete.

Quartzite is more ubiquitous in other areas of western New South Wales, occurring in the Barrier Range and the Scopes Range. The Scopes Range quartzite occurs as a pebble/boulder conglomerate bed running along the eastern edge of the range (Mills and Buckley 2000).

Silcrete

Silcrete is a hard-indurated rock with a matrix composed mostly of quartz clasts ranging in size from grains to boulders cemented by microcrystalline silica (Hutton *et al.* 1978:23). Silcrete is a very common element of Australian arid zone landscapes, occurring in the form of duricrusts and the residual capping of mesas and plateaux (Langford-Smith 1978:1). It also occurs as gibbers, which have resulted from silcrete breaking down through the processes of undermining and slumping (Hutton *et al.* 1978:20). Silcrete occurs widely across western New South Wales, but is generally less common in the Barrier Range than in areas to the north and east of Broken Hill. Occasionally, coarser grained silcretes have been misidentified as quartzite and finer grained silcretes as chert (Sullivan and Simmons 1979).

Doelman *et al.* (2001) characterised the silcretes from the Stud Creek location according to the proportion of microcrystalline matrix and clast size. A similar approach is employed here to characterise the silcretes represented in the Pine Point-Langwell stone artefact assemblages. Four main categories are defined:

- Amorphous – dominated by a microcrystalline matrix with no visible clasts
- Clast – medium to fine grained clasts
- Coarse – coarse grained clasts, often >5mm in diameter
- Matrix dominated – dominated by a microcrystalline matrix but contains scattered angular quartz clasts of varying sizes

In the following analyses these four silcrete categories are combined into two categories based on the overall character and organization of clasts within the matrix. These are:

- Clast silcrete – clast and coarse material
- Non-clast silcrete – amorphous and matrix dominated material

In other regions of western New South Wales, e.g., in the Stud Creek valley, these categories largely reflect the form in which the two types of silcrete are available (Doelman *et al.* 2001). Non-clast silcrete most commonly occurred as outcrop, while gibber pavements were predominantly composed of clast material, classified as medium-grained (Doelman *et al.* 2001:24). The investigation of silcrete outcrops within the vicinity of the Pine Point-Langwell study area indicates that the distribution and form of silcretes is different to that noted by Doelman *et al.* (2001) for Stud Creek. However, the two silcrete types remain useful analytical categories because they may indicate that one particular type of silcrete has been derived from sources outside the study area.

Silcrete is confined to two relatively small areas (compared to the overall size of the study area) of Tertiary deposits. The main area of silcrete occurs as outcrops and less frequently as gibber pavements on a series of ridge tops located in the Sampson's Paddock hills, approximately eight kilometres northeast of the Rantyga-Pine Creek confluence. A total of five outcrops with evidence of stone reduction (SQ1-5) and one major outcrop with limited reduction (SO1) were recorded. A smaller area of silcrete outcrops on the side and top of several low hills is located in the Enmore Hills on Sunnydale Station, approximately six kilometres northwest of the Rantyga-Pine Creek confluence. One outcrop with evidence of stone reduction was recorded (SQ6). The record of accumulation at silcrete outcrops may be much longer than that at the locations along the creeks, since these are residual surfaces dominated by slow erosion, and not subject to the types of geomorphic processes that obscure or destroy deposits in other areas of the landscape.

There are numerous small exposures of silcrete marked on the Redan and Balaclava 1:25,000 geological maps immediately to the north of the Pine Point-Langwell study area. These outcrops were not investigated for this study, however it is likely that at least some of them served as raw material sources for artefact production.

To gain some idea of the character of the quarry assemblages, a total of 4221 artefacts were recorded over a 292 square metre area at SQ1. The proportion of raw materials at SQ1 presented in Table 7.1 indicates that the assemblage is dominated by clast silcrete (>96%) by all three measures of raw material abundance. Non-clast silcrete, quartz and quartzite are the other materials represented in the quarry assemblage. Visual inspection of the other silcrete outcrops indicates a similar pattern of raw material representation. Non-clast silcrete is very uncommon at all the outcrops.

Cortex is common on artefacts at all the outcrops and is typically light brown in colour and abrasive to touch. The percentage of clast silcrete non-cortical and cortical complete flakes and cores at SQ1 is presented in Table 7.2. Almost 36% of complete flakes retain some cortex while 71.3% of cores retain cortex.

Table 7.1 Raw material proportion and percentage (in brackets) calculated by number, MNF and volume for the SQ1 assemblage.

Material	Number	MNF	Volume
Clast	4079 (96.6)	2195 (97.2)	91517155 (96)
Non-clast	73 (1.7)	39 (1.7)	659907 (0.7)
Quartz	52 (1.2)	17 (0.8)	2259681 (2.4)
Quartzite	17 (0.4)	7 (0.3)	856858 (0.9)
Total (Number)	4221 (100)	2258 (100)	95293601 (100)

Table 7.2 Frequency and percentage (in brackets) of clast silcrete non-cortical and cortical complete flakes and cores at SQ1. This indicates that cortex was recorded on the majority of clast silcrete cores and a large proportion of clast silcrete complete flakes in the quarry assemblage.

Category	Complete Flake	Core
Non-cortical	1064 (64.4)	31 (28.7)
Cortical	587 (35.6)	77 (71.3)
Total	1651 (100)	108 (100)

In addition to the silcrete outcrops, isolated silcrete gibbers were noted in the quartz gibber pavements that cover much of the Enmore Hills. As with quartz and quartzite gibbers, a smooth rounded cortex covers the external surface of silcrete gibber nodules and is readily distinguishable from the more abrasive cortex found on outcrop silcrete.

Summary

Despite being situated on the junction of the Barrier Range and the sandplain extending to the Darling River, the Pine Point-Langwell area contains a range of potential raw material sources suitable for the production of flaked stone artefacts. Materials suitable for ground stone artefacts, however, are absent from the area. Raw material sources have a restricted distribution and are concentrated in the north of the study area near the foothills of the Barrier Range. In the absence of the detailed mineral characterisation of raw material sources and the materials in the lithic assemblages (McBryde 1984, Sheppard 1997), it can only be assumed that some of the lithic materials in the assemblages, particularly quartz and clast silcrete were obtained from local sources.

Quartz occurs as fist sized gibber nodules and gravels within the immediate vicinity of all four lithic assemblages and is regarded as a local material. Clast silcrete sources do not occur within the immediate vicinity of any of the assemblages. KZ1 is situated closest to silcrete sources, approximately three kilometres from the silcrete quarries on Pine Point Station, while the other assemblages vary between five and seven kilometres distant (Figure 7.1). Clast silcrete is considered to be an imported material and the relative cost in terms of time and energy expenditure obtaining this material, even if procurement was embedded in other activities (see Binford 1979), is likely to be greater than that of quartz. Non-clast silcrete is not as readily available within the study area. Large proportions of this material are likely to have been obtained from sources situated outside the Pine Point-Langwell area. The cost of obtaining non-clast silcrete is likely to be greater than that of either clast silcrete or quartz.

Raw Material Representation and Characteristics

The potential procurement strategies by which raw materials are incorporated into assemblages are numerous. Two prominent strategies, direct and indirect (referred to as embedded) procurement, have dominated discussions of raw material economy in Australian arid zone archaeology (e.g., Binford and O'Connell 1984, Gould and Saggers 1985, O'Connell 1977). Discussions of trade have generally been confined to materials required for the manufacture of axes or seed grinding equipment (e.g., McBride 1997). Beyond investigating procurement strategies, the identification of which is usually ambiguous, quantifying the proportion of local and non-local raw materials in assemblages is important for understanding the parameters of lithic resource use.

Raw Material Proportion

The proportion of raw materials for each of the Pine Point-Langwell assemblages is displayed in Tables 7.3 to 7.5. Raw material proportions were calculated by the techniques discussed in Chapter Five. In general, both the minimum number of flakes and the number of individual pieces show a similar result for all of the assemblages, with quartz dominating the assemblages. However when calculated by volume the proportion of other raw materials generally increases relative to quartz. This can be explained by the high proportion of ground stone artefacts, which tend to be largely made of raw materials in the other category. The MNF and number measures are used as the primary indicators of raw material representation.

There are clear differences in the proportion of raw materials between the two CN and the two KZ assemblages. The MNF and the number of individual pieces indicate that all four assemblages are dominated by quartz, however the proportion of quartz in the CN assemblages is considerably greater than that for the KZ assemblages. Approximately 80% of raw materials in the CN1 and CN3 assemblages are composed of quartz. The two types of silcrete combined account for approximately 20% of raw materials in both CN1 and CN3, but silcrete is predominantly clast, with a slightly greater proportion of non-clast silcrete at CN1 (4%) compared to CN3 (3%).

The KZ1 and KZ2 assemblages demonstrate a different pattern of raw material representation to the CN1 and CN3 assemblages. By the MNF measure, quartz accounts for approximately 55% of raw materials at both KZ1 and KZ2, with this increasing to 60% when calculated by number. The two types of silcrete combined account for approximately 42% in both assemblages when calculated by MNF, but this decreases to between 37% and 38% when calculated by number. Therefore, the proportion of silcrete in the KZ assemblages is considerably greater than that for the CN assemblages. Again silcrete is dominated by clast material, accounting for approximately 35% at KZ1 and 38% at KZ2 when calculated by MNF. The proportion of non-clast silcrete is greater at KZ1 (6-7%) compared to KZ2 (4%).

Although both quartzite and crystal quartz occur naturally within the study area, these form only a minor amount of the overall raw material proportion. Quartzite is included in the others raw material category. The others category for each assemblage also includes chert, hornfels, ironstone, sandstone and schist. Sources for these materials are unknown, although sandstone makes up much of the Scopes Range to the northeast and schist occurs in the Barrier Range around Broken Hill.

Raw Material Quality

In the absence of mechanical fracture studies, the assessment of raw material quality is qualitative, with fracture characteristics inferred on the basis of texture

Table 7.3 Raw material by minimum number of flakes (MNF) for the four assemblages (percentage in brackets).

Material	CN1	CN3	KZ1	KZ2
Clast	666 (14.9)	367 (13.8)	544 (31.6)	2103 (36.9)
Non-clast	238 (5.3)	107 (4)	159 (9.2)	330 (5.8)
Quartz	3529 (78.8)	2133 (80.3)	967 (56.1)	3196 (56)
Other	45 (1)	49 (1.8)	54 (3.1)	74 (1.3)
Total (Number)	4478 (100)	2656 (100)	1724 (100)	5703 (100)

Table 7.4 Raw material by number of individual pieces for the four assemblages (percentage in brackets).

Material	CN1	CN3	KZ1	KZ2
Clast	1091 (12.4)	544 (11.1)	936 (27.4)	3526 (31.5)
Crystal Quartz	58 (0.7)	48 (1)	41 (1.2)	68 (0.6)
Non-clast	347 (3.9)	145 (3)	248 (7.3)	575 (5.1)
Quartz	7228 (82.2)	4107 (83.7)	2135 (62.4)	6895 (61.6)
Other	64 (0.7)	60 (1.2)	59 (1.7)	128 (1.1)
Total (Number)	8788 (100)	4904 (100)	3419 (100)	11192 (100)

Table 7.5 Raw material by volume (mm^2) for the four assemblages (percentage in brackets).

Material	CN1	CN3	KZ1	KZ2
Clast	16268761 (13.5)	7273643 (10.5)	13762712 (29.2)	41370556 (32.7)
Non-clast	2360261 (2)	919284 (1.3)	3356468 (7.1)	3191251(2.5)
Quartz	96114763 (80)	57949209 (83.4)	26229594 (55.7)	67350084 (53.2)
Other	5332306 (4.4)	3311571 (4.8)	3711478 (7.9)	14755882 (11.6)
Total (Volume)	120076091 (100)	69453707 (100)	47060252 (100)	126667773 (100)

and clast composition. A high quality raw material possesses little or no crystalline (clast) structure and contains few impurities and internal flaws (Brantingham *et al.* 2000:257). Elston (1990) argues that plasticity (workability) is an important attribute of raw material quality. Plasticity reflects the composition of the matrix. High quality raw materials are easy to control, produce flakes with sharp edges and require no pre-treatment to improve their flaking properties (Elston 1990:156-157). These criteria may be used to rank the quality of the three major raw material categories represented in the study assemblages. The matrix of non-clast silcrete comprises a non-crystalline structure and exhibits few internal flaws or impurities. Non-clast silcrete produces flakes with sharp edges and is considered a high quality raw material. Clast silcrete has a crystalline-based structure of various sized clasts (quartz grains), but exhibits few internal flaws or impurities. Clast silcrete flakes have durable sharp edges, but overall plasticity is not as high as that of non-clast silcrete. Quartz is dominated by internal flaws and has the least plasticity of the three materials.

The three major raw materials can be ranked from high plasticity (non-clast silcrete) to low plasticity (quartz). Non-clast silcrete is scarce and it is expected that efforts were made to conserve this material. Clast silcrete is abundant, but with geographically restricted sources, and it is expected that this material was extensively used. Quartz is locally abundant, but of the lowest utility and may be expected to be expediently worked.

Nodule Type

Two distinctive types of cortex are associated with raw material sources in the study area. Outcrop sources have an abrasive cortex referred to as non-gibber, while the smooth cortex found on gravels and weathered stone pavements is classified as gibber. These were easily distinguished in the study area. Calculating the non-gibber cortex to gibber cortex ratio for complete artefacts (flakes, tools and cores combined) provides an insight into the types of lithic sources that were exploited.

Table 7.6 Non-gibber to gibber cortex ratio for complete flakes, complete tools and cores combined. A ratio less than one indicate the exploitation of predominantly outcrop sources. Clast silcrete was primarily obtained from outcrop sources, while non-clast silcrete at CN1 and CN3 was mostly obtained from gibber sources. Quartz was obtained from gibber sources for all the assemblages.

Material	Assemblage	Gibber Cortex	Non-gibber Cortex	Ratio of Non-gibber to gibber cortex
Clast	CN1	26	29	0.9
	CN3	14	14	1
	KZ1	11	35	0.3
	KZ2	48	160	0.3
Non-clast	CN1	15	11	1.4
	CN3	5	2	2.5
	KZ1	1	11	0.1
	KZ2	19	20	0.9
Quartz	CN1	2048	3	682.7
	CN3	1314	1	1314
	KZ1	359	16	22.4
	KZ2	1593	7	227.6

The results for this ratio are presented in Table 7.6. The ratio for clast silcrete is less than one in the CN1, KZ1, and KZ2 assemblages, suggesting that nodules were more likely to have been obtained from outcrops rather than stone pavements or gravels. This is not true for CN3, where the ratio is one, suggesting that there is little difference between the proportion of clast silcrete nodules obtained form outcrops and gravels/stone pavements in this assemblage. The results for CN1 and CN3 suggest that non-clast silcrete was predominantly obtained from gravels and stone pavements. This is not true of KZ1 and KZ2, where the ratios are less than one, suggesting that non-clast silcrete was predominantly obtained from outcrop sources, although the ratio is only slightly less than one for KZ2. Quartz shows the clearest exploitation of gravel/stone pavement sources and as predicted there is very little evidence for the reduction of outcrop quartz. The ratio is especially high for CN1 and CN3, which are situated adjacent to gravel bars in Pine Creek. For the three major raw materials, only specific worked sources of clast silcrete could be defined.

Nodule Form

The form in which nodules were transported to the locations can be investigated through a comparison of the proportion of flakes with different amounts of cortex. If initial nodule decortification was undertaken at other locations, the proportion of complete flakes with large amounts of cortex will be low.

The majority of complete flakes in each assemblage for clast silcrete and non-clast silcrete do not have cortex. It is clear from Tables 7.7 and 7.8 that the percentage of cortical complete flakes in the four assemblages is low (<10%) for both silcretes. The majority of cortical complete clast and non-clast silcrete flakes have 1-50% cortex, with a much smaller proportion of those with 50-99% and complete cortex. From this it can be inferred that much of the initial reduction of both types of silcrete nodules occurred at other locations.

Although it is not possible to demonstrate a direct link between any of the worked silcrete outcrops and the assemblages recorded along the creeks, data from the SQ1 clast silcrete quarry may provide further information on the form in which clast nodules were transported from source areas. Table 7.9 presents the proportion of complete clast silcrete flakes with different amounts of cortex. These indicate that approximately 39% of complete flakes retain cortex, of which just below half have greater than 50% cortex. As predicted, the results are clearly different to those for the four creek assemblages and suggest on-source nodule reduction indicating that at least some material was removed from SQ1 as partially decortified nodules. This is a long-term pattern in raw material exploitation and demonstrates variation in core reduction strategies across the study area. Importantly, the pattern identified at the quarry is not taken to represent a direct link in the reduction process from source to creek assemblages. The chronological context of quarry exploitation is unknown due to the lack of heat retainer hearths, but the residual surfaces upon which the quarries occur are certainly as old as the surfaces in the creek valleys.

The opposite pattern is true for quartz (Table 7.10), where approximately 55% of complete flakes in the CN1 and CN3 assemblages, 50% in the KZ2 assemblage and 40% in the KZ1 assemblage have cortex. Additionally, the proportion of quartz complete flakes with 50-99% and complete cortex

Table 7.7 Frequency and percentage (in brackets) of clast silcrete complete flakes with different amounts of cortex. The proportion of cortical complete flakes is low in each of the assemblages and flakes with low proportions of cortex dominate this. These results suggest that a significant amount of nodule decortification occurred elsewhere.

Cortex category	CN1	CN3	KZ1	KZ2
None	431 (92.5)	258 (92.1)	358 (89.9)	1336 (89.5)
1-50%	19 (4.1)	15 (5.4)	31 (7.8)	104 (7)
50-99%	13 (2.8)	6 (2.1)	9 (2.3)	50 (3.4)
Complete	3 (0.6)	1 (0.4)	0 (0)	2 (0.1)
Total	466 (100)	280 (100)	398 (100)	1492 (100)

Table 7.8 Frequency and percentage (in brackets) of non-clast silcrete complete flakes with different amounts of cortex. The proportion of cortical complete flakes is low in each of the assemblages and flakes with low proportions of cortex dominate this. These results suggest that a significant amount of nodule decortification occurred elsewhere.

Cortex category	CN1	CN3	KZ1	KZ2
None	139 (90.8)	67 (91.8)	89 (89)	189 (88.3)
1-50%	12 (7.8)	3 (4.1)	6 (6)	18 (8.4)
50-99%	2 (1.3)	3 (4.1)	5 (5)	6 (2.8)
Complete	0 (0)	0 (0)	0 (0)	1 (0.5)
Total	153 (100)	73 (100)	100 (100)	214 (100)

Table 7.9 Frequency and percentage (in brackets) of clast silcrete complete flakes from the SQ1 silcrete quarry with different amounts of cortex. This assemblage exhibits a long-term pattern of raw material exploitation different to that of the four creek assemblages. Cortical flakes account for a much larger proportion of flakes at SQ1, indicating on source reduction.

Cortex category	Frequency
Complete	11 (0.8)
50-99%	231 (17.3)
1-50%	276 (20.7)
None	815 (61.1)
Total	1333 (100)

Table 7.10 Frequency and percentage (in brackets) of quartz complete flakes with different amounts of cortex. Quartz exhibits a different pattern to both the silcretes. Cortical complete flakes are common and a large proportion has greater than 50 percent cortex. This indicates on site reduction of cortical nodules.

Cortex category	CN1	CN3	KZ1	KZ2
None	1212 (44.6)	698 (41.3)	413 (58)	1222 (50.4)
1-50%	823 (30.3)	572 (33.9)	189 (26.5)	815 (33.6)
50-99%	537 (19.7)	329 (19.5)	97 (13.6)	321 (13.2)
Complete	147 (5.4)	90 (5.3)	13 (1.8)	66 (2.7)
Total	2719 (100)	1689 (100)	712 (100)	2424 (100)

Table 7.11 Length of complete non-cortical and cortical complete flakes from the CN1 assemblage. Clast silcrete cortical complete flakes have the largest length followed by quartz and non-clast silcrete. For each of the raw materials cortical flakes are on average longer than non-cortical flakes. Clast t=1.86, *df*=464, p=0.063, non-clast t=-1.92, *df*=14.04, p=0.076, quartz t=8.52, *df*=2709.98, p=<0.001.

Raw Material	Condition	*N*	Mean (mm)	St. Dev (mm)
Clast silcrete	Non cortical	431	26.6	11.2
	Cortical	35	30.2	11
Non-clast silcrete	Non cortical	139	20.7	5.1
	Cortical	14	25	8.2
Quartz	Non cortical	1212	23.9	7.1
	Cortical	1507	26.4	8.3

Table 7.12 Length of complete non-cortical and cortical complete flakes from the CN3 assemblage. Clast silcrete cortical complete flakes have the largest length followed by quartz and non-clast silcrete. For clast silcrete and quartz cortical flakes are on average longer than non-cortical flakes, however this is not true of non-clast silcrete. Clast t=1.73, *df*=22.67, p=0.097, non-clast t=-0.537, *df*=71, p=0.593, quartz t=7.01, *df*=1621.25, p=<0.001.

Raw Material	Condition	*N*	Mean (mm)	St. Dev (mm)
Clast silcrete	Non cortical	258	25.9	11.4
	Cortical	22	32.2	16.9
Non-clast silcrete	Non cortical	67	21.6	6.2
	Cortical	6	20.2	4.1
Quartz	Non cortical	698	23.9	7.1
	Cortical	991	26.2	8.3

is considerably greater than that for both types of silcrete. This suggests that quartz nodules were transported to the locations as cortical nodules, most likely from sources within the immediate vicinity of the place of reduction.

Nodule Size

Estimating original nodule size is difficult. This is especially true when the entire core reduction sequence is not represented, as is the case for clast and non-clast silcrete in the four assemblages. Calculating the average size of complete flakes that retain high proportions of dorsal cortex is one method of investigating the effects of nodule size on core reduction. Cortical flakes are usually, but not always, removed during the early stages of reduction (Baulmer 1988); therefore they provide a relative indication of original nodule size.

The mean length of non-cortical complete flakes and cortical complete flakes for the four assemblages (Tables 7.11–7.14) indicates that clast silcrete nodules are on average larger than either non-clast silcrete or quartz nodules and this difference is significant (Table 7.15). There is no significant difference in the size of quartz cortical complete flakes and non-clast silcrete cortical complete flakes in each assemblage.

Comparison of non-cortical complete flake length to that of cortical complete flakes indicates that for all raw materials in the four assemblages, except non-clast silcrete at CN3, the length of cortical flakes is significantly greater than that for non-cortical flakes. For non-clast silcrete the difference is only significant at CN1. The significant results suggest that some of the nodules reduced at each location were covered in cortex. If this were not the case, the size difference between non-cortical and cortical flakes would not be evident. The anomalous result for non-clast silcrete for CN3 may reflect the effect of outliers because of the small number of cortical flakes (six) in the assemblage. It appears that non-clast silcrete nodules containing cortex were not reduced at CN3 as often as at the three other locations. Complete cortical flake length indicates that the only statistically significant difference was between quartz at KZ2 and that at CN1 and CN3 (Table 7.16). This further suggests that quartz nodules at KZ2 were generally smaller.

Discussion

The four Pine Point-Langwell assemblages are dominated by locally available quartz, with smaller amounts of imported silcrete accounting for 20% of raw materials in the CN1 and CN3 assemblages and approximately 40%

Table 7.13 Length of complete non-cortical and cortical complete flakes from the KZ1 assemblage. Clast silcrete cortical complete flakes have the largest length followed by quartz and non-clast silcrete. For each of the raw materials cortical flakes are on average longer than non-cortical flakes. Clast t=2.58, *df*=44.15, p=0.13, non-clast t=.263, *df*=13.37, p=0.796, quartz t=3.81, *df*=541.33, p=<0.001.

Raw Material	Condition	*N*	Mean (mm)	St. Dev (mm)
Clast silcrete	Non cortical	358	27.9	10.5
	Cortical	40	33.8	13.9
Non-clast silcrete	Non cortical	89	24.4	7.6
	Cortical	11	25	6.7
Quartz	Non cortical	413	23.4	7
	Cortical	299	25.8	9

Table 7.14 Length of complete non-cortical and cortical complete flakes from the KZ2 assemblage. Clast silcrete cortical complete flakes have the largest length followed by quartz and non-clast silcrete. For each of the raw materials cortical flakes are on average longer than non-cortical flakes. Clast t=6.07, *df*=180.04, p=<0.001, non-clast t=.822, *df*=212, p=0.412, quartz t=7.81, *df*=2318.58, p=<0.001.

Raw Material	Condition	*N*	Mean (mm)	St. Dev (mm)
Clast silcrete	Non cortical	1336	24.3	9
	Cortical	156	29.9	11
Non-clast silcrete	Non cortical	189	21.6	6.6
	Cortical	25	22.8	8.3
Quartz	Non cortical	1222	23.2	6.2
	Cortical	1202	25.4	7.6

Table 7.15 Results of least significance difference test (LSD) for complete cortical flake length between the three major raw material categories within the four assemblages. Only those results that returned a significant difference are displayed.

Assemblage	Material	Material	Mean difference	*p*
CN1	Clast	Non-clast	5.23	0.050
	Clast	Quartz	3.85	0.007
CN3	Clast	Non-clast	12.06	0.002
	Clast	Quartz	5.66	0.002
KZ1	Clast	Non-clast	8.78	0.008
	Clast	Quartz	7.96	<0.001
KZ2	Clast	Non-clast	7.01	<0.001
	Clast	Quartz	4.49	<0.001

Table 7.16 Results of least significance difference test (LSD) for complete cortical flake length between within raw material class between assemblages. Only those results that returned a significant difference are displayed.

Material	Assemblage	Assemblage	Mean difference	*p*
Quartz	CN1	KZ2	1.01	0.001
	CN3	KZ2	1.21	0.001

in the KZ1 and KZ2 assemblages. The relative proportion of raw materials in the Pine Point-Langwell assemblages suggest a distance decay relationship of decreased abundance with increased distance from source. Both the KZ1 and KZ2 assemblages are several kilometres closer to silcrete sources than the CN assemblages. Although a distance decay relationship for silcrete is interesting, it is not surprising (Odell 1996). Linear distance decay relationships have been noted in many arid zone stone artefact studies (e.g., Doelman *et al.* 2001, Hiscock 1985, O'Connell 1977). Barton (2001) investigated assemblage composition in arid western Queensland, arguing that raw material distribution was not an important factor influencing assemblage composition. However, in the Pine Point-Langwell study raw material distribution was clearly a significant influence on assemblage composition, even over short distances. Even though clast silcrete was imported into each location, there are numerous nearby sources. These are distances that some researchers working in Europe consider to be local (e.g., Montet-White 1991).

Many of the studies reviewed in Chapter Three indicate that raw materials are not always transported across landscapes in linear directions. Aspects of raw material conservation are manifested in the technological composition of lithic assemblages and cannot be evaluated through raw material proportions alone. Commonly referred to as economising behaviour, this may be evidenced by increased tool resharpening and more intensive core reduction.

Summary

There is considerable variation in the relative proportion of raw materials across the study area. Local raw materials dominate the assemblages, but there is a clear distance decay relationship evident. The assemblages can be divided into two broad groups. KZ1 and KZ2 have a much greater proportion of silcrete than CN1 and CN3, where locally available quartz is dominant. In a spatial context, this represents significant variation between the assemblages. From a temporal perspective, raw material availability in the study area is a constant, and it has served to structure a long-term accumulated pattern. Therefore, a degree of compositional variation will reflect this, but to what extent remains to be demonstrated in the next chapter.

Raw material distribution and availability has operated as a long-term structuring factor. At the same time the attributes of individual raw material types is a constraint on the flexibility of reduction strategies. These attributes transcend the discontinuities represented in the occupational chronologies. Other aspects of the environment may not be as stable over the long-term, and these may have significant implications not only for the temporal continuity of occupation, but also for place use histories. Behavioural variability can be expected to fluctuate in conjunction with environmental change. This can be explored in a detailed analysis of assemblage composition.

Chapter Eight: Raw Material Reduction and Assemblage Composition

Introduction

This chapter presents a detailed analysis of the composition of the Pine Point-Langwell assemblages aimed at establishing the degree and range of variability in raw material utilisation and reduction between the assemblages. As described in Chapter Three, a series of economic analyses are used to investigate the relationship between artefact reduction, discard and life history. Radiocarbon determinations from heat retainer hearths indicate that the locations were occupied on multiple occasions. Therefore, the aggregate assemblages represent artefacts discarded during occupations that may differ in character (temporal, spatial and functional). The formation of the assemblages demonstrates that it is not valid to view these as single function sets.

The structure of the lithic analysis employed here closely follows that of North American approaches, especially those applied to the study of Palaeolithic (e.g., Dibble 1987, 1995a, Kuhn 1995) and Palaeo-Indian assemblages (e.g., Shott 1997). These studies apply quantitative methods to the investigation of raw material utilisation and reduction in flake-based stone artefact assemblages. Aspects of these approaches have recently been applied in Australia to the investigation of flake tool resharpening (e.g., Clarkson 2002, Hiscock and Attenbrow 2003) and raw material reduction (e.g., Holdaway 2000, Holdaway *et al.* 2004, Shiner *et al.* 2005).

The analysis is presented in three main parts, beginning with the investigation of core reduction techniques through to an examination of the representation of core forms. The second part of the chapter investigates the intensity of core reduction. Of particular interest is the nature of flake production, as core reduction becomes more intensive. The final part of the analysis examines the nature of tool production, maintenance and resharpening. The chapter concludes with a summary of the characteristics of the assemblages.

Raw Material Utilisation, Occupation Intensity and the Composition of Stone Artefact Assemblages

Several researchers (e.g., Dibble 1995b, Kuhn 1995, Veth 1993) suggest that the intensity of raw material utilisation increases with the intensity of occupation. This relationship is most clearly expressed in assemblage composition. Assemblages from intensively occupied locations will exhibit greater degrees of raw material consumption, higher proportions of long use-life tools and a more extensive reliance on local raw materials. These factors together can be used to identify the places that experienced the greatest degree of use through time. Kuhn (1995) for example applied these criteria to the investigation of the Middle Palaeolithic in Italy to identify periods of more intensive rockshelter occupation. Holdaway *et al.* (2004) and Shiner (1999) applied these measures to rangeland assemblages in western New South Wales. Other applications (e.g., Shiner *et al.* 2005) indicate that relationships between greater raw material utilisation and intensity of occupation are not always clear. Some assemblages demonstrate variability that is not easily classified into clearly defined relationships between raw material economy and occupation intensity.

Kuhn's model is essentially a synchronic one in which historic variability is reduced to a minor role. The result is a uniform set of relationships between raw material utilisation and occupation intensity operating through time. Kuhn argues that there is a consistent relationship between technological strategies and models of raw material provisioning. He argues that transport costs and raw materials are limiting variables, as noted by others (e.g., Elston 1990). In the case of the assemblages from coastal Italy, he argues that assemblage composition is closely correlated with the procurement of game. The result of this relationship is one of stability in stone artefact manufacture over a period of 60,000 to 70,000 years (Kuhn 1995:173). Essentially people must be behaving in roughly the same manner over long periods of time for a uniform pattern to form.

Kuhn argues for a long-term relationship between provisioning and technology in the Middle Palaeolithic. This contrasts to the record of modern humans, where considerable variability in assemblage composition is evident across time and space. Some factors such as the distribution and characteristics of raw materials will condition and constrain historical variability, but these are not the only factors that control the overall composition of assemblages. In arid Australia, the likelihood that people occupied locations the same way throughout the period over which assemblages have accumulated is low, especially in environments that experience significant short-term variability and change over the long-term. Important to this point is the nature of the assemblages themselves as artificial combinations of artefacts from many separate episodes of occupation.

With these factors in mind, the following analysis pursues an investigation of historical variability through an investigation of the degree to which a uniform pattern in assemblage composition is represented in the Pine Point-Langwell assemblages. The notion of raw material characteristics as constraints on variability was addressed in Chapter Seven, and it is now possible to assess the degree of historical variability between the assemblages free from purely economic considerations. If the Pine Point-Langwell assemblages reflect behavioural continuity in place use, a high degree of consistency is expected in the various measures of the intensity of raw material utilisation and assemblage composition. On the other hand, intra assemblage variability and inconsistencies would suggest a level of variation between the occupation episodes. This arises from variable occupation histories, not necessarily in terms of the activities, but rather from the duration, intensity and frequency of occupations.

Core Reduction Techniques

Core reduction strategies are a fundamental influence on the composition of stone artefact assemblages (Baumler 1988:255). They refer to the methods by which flakes are removed from cores. These reflect many factors, including the character of available raw materials (Baumler 1988:260). Calculating the proportion of different core forms provides an insight into the nature of core reduction strategies.

Raw Material Quality, Nodule Type, Form and Size

In Chapter Seven it was demonstrated that the four assemblages contain the same types of raw material, but in different proportions. While much of this pattern reflects distance decay relationships, some aspects may relate to decisions based on raw material quality or the overall occupation histories of the locations. Choices based on raw material quality may be a major factor influencing potential variability between assemblages. This can be assessed through a consideration of raw material and artefact manufacture. Studies in other areas of western New South Wales (e.g., Shiner *et al.* 2005) indicate significant variability in core reduction and tool manufacture between quartz, clast silcrete and non-clast silcrete. In addition to raw material quality, differences in artefact manufacture between these materials may reflect the characteristics of nodule type, form and size.

As demonstrated in the previous chapter, nodule type, form, and size are similar within single raw material categories when compared between the Pine Point-Langwell assemblages. There are considerable differences between the three major raw materials, and these are largely mirrored in each of the Pine Point-Langwell assemblages. Quartz nodules were almost exclusively derived from gibber pavements and gravel bars. These are covered in a smooth and rounded cortex and are considerably smaller in size than clast silcrete nodules. Analyses of cortical flake length in the previous chapter indicated that clast silcrete nodules are the largest of the three raw materials, and were mainly obtained from outcrops. Comparison of clast cortical flake length between the four assemblages indicated that nodules were largest in the KZ1 assemblage, which is also the closest to the clast silcrete outcrops in the foothills above Rantyga Creek. This is characteristic of a distance decay relationship. Non-clast silcrete nodules are the smallest raw material, and these were obtained from both gibber pavements and outcrops. The absence of non-clast silcrete sources in the study area suggests that the majority of this material was brought into the area from sources located elsewhere.

There are a number of shared consistencies between the assemblages. Quartz nodules were obtained from similar sources for each assemblage, as was clast silcrete. While the exact location of the sources is not known, the potential sources of these materials have been identified in the study area. Similar raw material characteristics shared between the assemblages may constrain the level of variability in reduction and assemblage composition across the study area. This may be seen in technological measures of assemblage reduction. For instance, quartz nodules are similar in size in each of the assemblages and therefore may demonstrate a similar level of reduction independent of other factors such as mobility and occupation duration. With the potential influence of raw material characteristics accounted for, an investigation of variability arising from other factors reflecting the character of the occupations can be undertaken.

Core Form

Core form is indicative of the techniques used to reduce nodules. For example, non-specialised core forms with platforms flaked from only one direction (unifacial) and with only one or two negative flake scars (test), suggest non-intensive core reduction strategies. Those with platforms flaked from two or more directions (bifacial and multiple forms) indicate core rotation and a concern with extending the reduction life of nodules.

The proportion of clast silcrete core forms in each assemblage is presented in Table 8.1. The proportion of individual core types per assemblage is calculated as a proportion of the overall total number of cores in each assemblage. Eight different forms are represented, with eight forms at KZ2, five at KZ1, four at CN1 and three at CN3. Specialised forms comprising microblade, radial and flake-blank are uncommon across each assemblage. Microblade, nuclear-tool and radial core forms were only recorded at KZ2, while flake-blank forms were confined to KZ1 and KZ2. Proportions of test core forms vary from 9.4% at CN1 followed by 3% at KZ1 and 2.6% at KZ2 and are not represented at CN3.

Unifacial cores are the single most common clast silcrete core form in all of the assemblages, varying from 75% at CN1, 71.5% at CN3, 54.8% at KZ2 and 48.5% at KZ1. At KZ1 the combined proportion of rotated forms (45.5%) is smaller than that of unifacial forms (48.5%), although this is still a relatively high proportion of rotated forms and is much greater than that for the other three assemblages. The proportion of rotated forms decreases from 31.3% at KZ2 to 28.5% at CN3 and 15.6% at CN1.

Table 8.1 Frequency and proportion (in brackets) of clast silcrete core types per assemblage. Unifacial cores are the most common form. The combined proportion of rotated forms (bifacial and multiple) is highest at KZ1.

Core type	CN1	CN3	KZ1	KZ2
Bifacial	4 (12.5)	2 (9.5)	6 (18.2)	31 (27)
Flake Blank			1 (3)	8 (7)
Microblade				1 (0.9)
Multiple	1 (3.1)	4 (19)	9 (27.3)	5 (4.3)
Nucleartool				1 (0.9)
Radial				3 (2.6)
Test	3 (9.4)		1 (3)	3 (2.6)
Unifacial	24 (75)	15 (71.5)	16 (48.5)	63 (54.8)
Total	32 (100)	21 (100)	33 (100)	115 (100)

Table 8.2 Frequency and proportion (in brackets)of non-clast silcrete core types per assemblage. Unifacial cores are the most common form in all assemblages except KZ1, where the proportion of bifacial forms is greater. Again the combined proportion of rotated forms is highest at KZ1.

Core type	CN1	CN3	KZ1	KZ2
Bifacial	2 (20)		4 (36.4)	5 (21.7)
Bipolar	1 (10)			
Flake Blank			1 (9.1)	2 (8.7)
Multiple	1 (10)	1 (20)	2 (18.2)	1 (4.3)
Nucleartool				2 (8.7)
Radial			1 (9.1)	1 (4.3)
Test			1 (9.1)	
Unifacial	6 (60)	4 (80)	2 (18.2)	12 (52.2)
Total	10 (100)	5 (100)	11 (100)	23 (100)

The proportion of non-clast silcrete core forms is presented in Table 8.2. There are numerically fewer non-clast cores in all of the assemblages compared to the number of quartz and clast silcrete cores. A total of eight different forms are represented across all assemblages, with six forms represented in the KZ1 and KZ2 assemblages, four in CN1 and two in CN3. Specialised core forms comprising bipolar, flake-blank and radial are uncommon. One bipolar core was recorded at CN1, while one flake-blank core was recorded at KZ1 and two at KZ2. One radial core was recorded at both KZ1 and KZ2. Test core forms were also rare, with only one test core recorded at KZ1. Two nuclear-tool forms were recorded at KZ2.

Unifacial cores are the single most common non-clast core form in all of the assemblages except KZ1, varying from 80% at CN3, followed by 60% at CN1, 52.2% at KZ2 and 18.2% at KZ1. At KZ1 the proportion of bifacial cores is twice as great as unifacial cores (36.4% compared to 18.2%) and the combined proportion of rotated core forms is considerably higher than that of unifacial forms at KZ1 (54.6% compared to 18.2%). This pattern does not follow at CN1, CN3 and KZ2. These assemblages show similar patterns to those identified for clast silcrete.

The proportion of core forms for quartz is presented in Table 8.3. A total of nine different forms are represented across all assemblages, with eight different forms represented in all of the assemblages. Specialised forms including microblade, radial and flake-blank are uncommon. Microblade forms are confined to KZ2, while radial cores only occur at CN1, CN3 and KZ1. Bipolar cores are also uncommon in all assemblages. Nuclear-tool forms account for less than 5% of quartz core forms in all of the assemblages. Test core forms are relatively uncommon across all assemblages. The highest proportion of test cores occurs at CN1 (8.5%) and CN3 (6.1%).

Table 8.3 Frequency and proportion (in brackets) of quartz core types per assemblage. Unifacial cores are the most common form, and the proportion of rotated forms is again highest at KZ1.

Core type	CN1	CN3	KZ1	KZ2
Bifacial	92 (18.3)	49 (16.7)	37 (26.6)	118 (28.2)
Bipolar	6 (1.2)	2 (0.7)	6 (4.3)	19 (4.5)
Flake Blank	5 (1)	8 (2.7)	1 (0.7)	11 (2.6)
Microblade				2 (0.5)
Multiple	46 (9.1)	19 (6.5)	12 (8.6)	11 (2.6)
Nucleartool	19 (3.8)	3 (1)	6 (4.3)	6 (1.4)
Radial	2 (0.4)	1 (0.3)	1 (0.7)	
Test	43 (8.5)	18 (6.1)	5 (3.6)	19 (4.5)
Unifacial	290 (57.7)	194 (66)	71 (51.1)	233 (55.6)
Total	503 (100)	294 (100)	139 (100)	419 (100)

Unifacial cores are the most common quartz core form in all of the assemblages, varying from 51.1% at KZ1 to over 66% at CN3. Again the proportion of rotated core forms (bifacial and multiple combined) is highest at KZ1 (34.2%). The low proportion of rotated core forms indicates that core rotation was not a priority for quartz reduction.

Some generalised patterns in core reduction can be inferred from the quantification of core form. Specialised core forms are rare in all three raw material classes in each of the assemblages. Core rotation is relatively uncommon, especially on clast silcrete, however there is considerable variability in the proportion of rotated core forms between assemblages and raw materials. KZ1 has the highest proportion of rotated forms for each of the raw materials. This suggests that extending the reduction life of cores was an important feature of the reduction strategy at KZ1 compared to the other three assemblages. This result is somewhat unexpected, especially for clast silcrete, because KZ1 is located the closet of all the assemblages to silcrete outcrops. The CN1 and CN3 assemblages have the lowest proportions of rotated clast silcrete and non-clast silcrete core forms, a result that does not follow a simple linear distance decay model. Bipolar core forms are rare, especially for quartz where these might be expected. This suggests quartz cores were discarded prior to reaching the inertia threshold requiring further reduction using bipolar techniques. Variability in core form proportions suggests that differences in the degree and form of core reduction and flake production exist between the assemblages.

Core Reduction Intensity

Many characteristics of stone artefact assemblages change in accordance with the degree of core reduction (Blades 1999:103, Roth and Dibble 1998:49). As core reduction becomes more intensive the number of flakes produced increases, flake and core size decreases and the proportion of cortical to non-cortical flakes and cores decreases. The degree to which cores are reduced can also reflect occupation duration. Longer duration occupation by less mobile groups may result in more intensive core reduction, producing more flakes per core on average (Dibble 1995a:354, Elston 1990:160). The analysis of core forms indicated that KZ1 had the greatest proportion of rotated forms for each of the raw material classes, suggesting that core reduction was more intensive in this assemblage compared to the other three assemblages. This hypothesis is tested in the following set of analyses.

MNF (Minimum Number of Flakes) to Core Ratio

The MNF to core ratio is the most basic measure of core reduction intensity. As core reduction proceeds, the number of flakes produced increases (Dibble 1995b:292). Therefore, assemblages that have high flake to core ratios indicate increased core reduction. Bipolar cores and flakes were not included in the quartz flake to core ratio because these represent a different technique of core reduction to unipolar reduction. This measure is based on the assumption that through time both cores and flakes are discarded and transported. Therefore not every single core and flake remains at the location at which it was produced, but it is still a useful measure of the intensity of core reduction.

Calculating this ratio for each raw material type (Table 8.4) indicates both similarities and differences between the assemblages. In all of the assemblages the ratio for quartz is lowest indicating that this local material was the least intensively reduced. Of the three major raw material types, the quartz ratio also shows the least variability between assemblages. KZ2 has the highest ratio with eight and CN1 the lowest with 7.1, while the ratio is 7.3 for both CN3 and KZ1.

The flake to core ratio for clast silcrete is lower than that for non-clast silcrete in the CN1, CN3 and KZ2 assemblages, but at KZ1 the clast value is higher than non-clast. The CN1 assemblage has the highest ratio for clast silcrete, followed by KZ2, CN3 and KZ1. In general the clast silcrete ratio is

Table 8.4 MNF (minimum number of flakes) to core ratio per raw material and assemblage. The value for quartz is lowest in each of the assemblages, while the value for non-clast silcrete is highest in each assemblage except KZ1 indicating that there are proportionally more flakes per core for non-clast silcrete compared to the other raw materials.

Material	CN1	CN3	KZ1	KZ2
Clast	20.8	17.5	16.5	18.3
Non-clast	23.7	21	14.5	25.4
Quartz	7.1	7.3	7.3	8

Table 8.5 Non-cortical to cortical complete flake ratio per raw material and assemblage. Quartz has the lowest ratio for each of the assemblages and at CN1 and CN3 there are proportionally more cortical flakes than non-cortical flakes. Clast silcrete has the highest ratio, a result that does not follow that of the MNF to core ratio.

Material	CN1	CN3	KZ1	KZ2
Clast	12.3	11.7	9	8.6
Non-clast	9.9	11.2	8.1	7.6
Quartz	0.8	0.2	1.4	1

Table 8.6 Non-cortical to cortical core ratio per raw material and assemblage. Quartz has the lowest ratio in each assemblages (except non-clast silcrete at CN3) indicating that there are more cortical than non-cortical cores. Again, proportionally more clast silcrete cores do not contain cortex than cores on the other two raw materials.

Material	CN1	CN3	KZ1	KZ2
Clast	2.6	3.2	3.7	2.1
Non-clast	1.5	0	2.7	1.3
Quartz	0.2	0.1	0.6	0.3

similar between the assemblages. The higher ratio at CN1 and the lower ratio at KZ1 are consistent with a distance decay relationship, but this is not the case for either the KZ2 or CN3 assemblages. The ratios for non-clast silcrete demonstrate a greater amount of variability between the assemblages than is the case for clast silcrete and quartz. KZ2 has the highest ratio for non-clast silcrete, followed by CN1 and CN3.

None of the assemblages demonstrate a clear pattern of intensive core reduction in all of the raw material categories. KZ1 has the lowest value for clast and non-clast silcrete and the second lowest for quartz, suggesting that despite the high representation of rotated core forms, the reduction of these was not very intensive. Non-intensive core reduction at KZ1 is representative of a closer proximity to major clast silcrete sources than is true for the other assemblages. Both CN1 and KZ2 have the highest ratios for clast and non-clast silcrete, indicating that core reduction was most intensive in these assemblages. In addition, KZ2 has the highest ratio for quartz. CN3 exhibits more intensive core reduction than KZ1, but this is not as intensive as either CN1 or KZ2.
There is little correlation between larger proportions of rotated core forms and increased core reduction in the assemblages. Either core rotation was not an important method of extending nodule reduction life, or there was little economic imperative for increasing the reduction life of cores.

Non-cortical to Cortical Complete Flake Ratio

Increased core reduction leads to a decrease in the proportion of artefacts with cortex, therefore higher ratios indicate more intensive core reduction (Holdaway *et al.* 2004). The MNF to core ratios suggested that non-clast silcrete was generally the most intensively worked material, thus it is expected that the non-cortical to cortical complete flake ratio will be highest for this material. Results for this ratio (Table 8.5) indicate that this is not the case, instead the ratio is highest for clast silcrete across all the assemblages. Non-clast silcrete has the second highest ratio, followed by quartz. The ratio is consistent for quartz and reflects the reduction of local cortical nodules. The same cannot be said for clast silcrete. Clearly clast silcrete, although locally available, was not utilised in the same way as local quartz. The KZ1 and KZ2 ratios for quartz are higher than the ratios for CN1 and CN3, suggesting more intensive reduction of quartz cores in the KZ assemblages, a hypothesis supported by the MNF to core ratio for KZ2, but not KZ1. The patterns emerging at KZ2 and to a lesser extent KZ1, cannot be attributed to differential raw material access. CN1 and CN3 show similar values for both types of silcrete. This result supports the relatively high values of the MNF to core ratio and suggests that silcrete reduction was intensive in these two assemblages. KZ1 and KZ2 have similar results for both types of silcrete. This result is surprising because the MNF to core ratio suggested that core reduction was more intensive at KZ2 than KZ1.

Non-cortical to Cortical Core Ratio

The non-cortical to cortical core ratio (Table 8.6) is based on the same principle as the non-cortical to cortical flake ratio.

Clast silcrete has the highest ratio for each of the assemblages, followed by non-clast silcrete, except for CN3, where the ratio could not be calculated because no cores with cortex were recorded. Quartz has the lowest ratio in all the assemblages, with the ratio not exceeding one, indicating that there are more cortical than non-cortical cores. Across raw materials the results for this ratio generally follow a similar pattern to that of the cortical to non-cortical complete flake ratio.

The highest ratios for the three raw material types all occur in the KZ1 assemblage. This is unexpected because the previous two analyses suggested that KZ1 cores are the least intensively worked, but it does fit with the high proportion of rotated core forms at KZ1, which suggest a greater likelihood of cortex removal resulting from the flaking of multiple surfaces. The results for the CN1 and KZ2 assemblages are also ambiguous. The MNF to core and the non-cortical to cortical complete flake ratios suggested that core reduction was intensive at CN1. This is not supported by the non-cortical to cortical core ratio, which indicates a low proportion of decortified cores at CN1. There is a similar result for KZ2, but the pattern is less clear. The MNF to core ratio was high at KZ2, but the non-cortical to cortical complete flake ratio low. The low proportion of decortified cores at KZ2 supports the non-cortical to cortical complete flake ratio, but is in disagreement with the MNF to core ratio. CN3 follows a pattern consistent with the non-cortical to cortical complete flake ratio.

Inconsistencies in the technological indices point to the variable nature of assemblage formation. Some of these inconsistencies may reflect the variable nature of the behavioural processes responsible for assemblage accumulation. These behaviours may include artefact recycling, and artefact removal as well as variable occupation spans, this will be further discussed below.

Core Size

As core reduction becomes more intensive core size will decrease. The mean maximum length, width and thickness of all cores (excluding bipolar forms) are calculated to investigate the relationship between core reduction and core size at discard.

Results in Tables 8.7 to 8.9 indicate that non-clast silcrete cores have the smallest average dimensions in three of the four assemblages. The exception is KZ1, where non-clast silcrete cores are on average larger than quartz cores. Quartz cores at CN1, CN3 and KZ2 have larger average dimensions than those of non-clast silcrete, but this difference is only significant for maximum length at CN1 (Table 8.10). Clast silcrete cores are the largest in each of the assemblages. This result is significant at CN1 and KZ2 (Tables 8.10 and 8.12), but for KZ1 the mean maximum dimensions are only significantly different between clast silcrete and quartz (Table 8.11). Results for the CN3 assemblage are not tabulated because the LSD test returned non-significant results.

Table 8.7 Mean maximum length, width and thickness of clast silcrete cores in the four assemblages. Maximum length $F=0.68$, $df\,3, 197$, $p=0.564$, maximum width $F=1.87$, $df\,3, 197$, $p=0.136$, maximum thickness $F=2.05$, $df\,3, 197$, $p=0.108$

Assemblage		Max. Length (mm)	Max. Width (mm)	Max. Thickness (mm)
CN1	Mean	60.3	50.1	35.1
	S.D.	24.8	22.1	12.2
	N	32	32	32
CN3	Mean	55.3	43.9	32.6
	S.D.	21	14.4	11.6
	N	21	21	21
KZ1	Mean	54.3	46.6	36.3
	S.D.	12.4	11.4	11
	N	33	33	33
KZ2	Mean	54.3	42.6	30.9
	S.D.	22.4	16.8	13.6
	N	115	115	115

Table 8.8 Mean maximum length, width and thickness of non-clast silcrete cores in the four assemblages. Maximum length F=2.20, *df* 3, 45, p=0.101, maximum width F=2.18, *df* 3, 45, p=0.103, maximum thickness F=2.25, *df* 3, 45, p=0.096

Assemblage		Max. Length (mm)	Max. Width (mm)	Max. Thickness (mm)
CN1	Mean	39.1	30.3	23.1
	S.D.	10.4	8.8	8
	N	10	10	10
CN3	Mean	38	29.6	25.6
	S.D.	4.9	7.6	4.9
	N	5	5	5
KZ1	Mean	50	42.5	29.5
	S.D.	29.2	27.5	16.8
	N	11	11	11
KZ2	Mean	34.7	28.9	20.3
	S.D.	9.8	8.6	6.3
	N	23	23	23

Table 8.9 Mean maximum length, width and thickness of quartz cores in the four assemblages. Maximum length F=9.33, *df* 3, 1318, p=<0.001, maximum width F=9.50, *df* 3, 1318, p=<0.001, maximum thickness F=10.99, *df* 3, 1318, p=<0.001

Assemblage		Max. Length (mm)	Max. Width (mm)	Max. Thickness (mm)
CN1	Mean	50.3	39.8	30.2
	S.D.	18.5	14.8	11
	N	497	497	497
CN3	Mean	50.9	39.9	30.2
	S.D.	20.6	15.1	10.7
	N	292	292	292
KZ1	Mean	45.2	36	27.5
	S.D.	19.5	15.2	10.8
	N	133	133	133
KZ2	Mean	44.9	35.4	26.6
	S.D.	17.5	13.1	10.2
	N	400	400	400

The KZ2 assemblage has the smallest mean values for each measurement in the three raw material categories. Results of the least significance difference test indicate that for clast silcrete the differences between the means in the four assemblages are generally not significant (Table 8.13). For non-clast silcrete, the mean dimensions of KZ2 cores are consistently significantly smaller than KZ1 cores. Quartz cores form two distinct size groups with both CN1 and CN3 cores being on average significantly larger than cores from KZ1 and KZ2.

Although the differences in mean core size between KZ2 and the other assemblages were not consistently significant, the smaller average size of cores does suggest that KZ2 cores are the most intensively reduced. This result may also reflect the size of the nodules transported to KZ2. Cortical flake length provides an indication of the original size of nodules. Analysis of complete cortical flake length in Chapter Seven indicated that KZ2 had the smallest average for both clast silcrete and quartz, and the second smallest for non-clast silcrete. This suggested that the nodules transported to KZ2 were on average smaller for clast silcrete and quartz, and relatively small for non-clast silcrete compared to nodules in the other assemblages. The small size of KZ2 cores may therefore reflect the extended reduction of smaller nodules. On its own this does not necessarily represent more intensive core reduction relative to other assemblages, because the nodules were smaller to begin with. If the MNF to core ratio is also considered, then there is evidence that non-clast silcrete cores were more intensively reduced.

Table 8.10 Results of least significance difference test (LSD) for cores on different raw materials in the CN1 assemblage. Only those results that returned a significant difference are displayed.

Variable	Material	Material	Mean difference	*p*
Maximum length	Clast	Non-clast	21.18	<0.001
	Clast	Quartz	10.01	<0.001
	Non-clast	Quartz	11.17	0.052
Maximum Width	Clast	Non-clast	19.83	<0.001
	Clast	Quartz	10.37	<0.001
Maximum Thickness	Clast	Non-clast	11.96	0.003
	Clast	Quartz	4.83	0.017
	Non-clast	Quartz	-7.14	0.043

Table 8.11 Results of least significance difference test (LSD) for cores on different raw materials in the KZ1 assemblage. Only those results that returned a significant difference are displayed.

Variable	Material	Material	Mean difference	*p*
Maximum Length	Clast	Quartz	9.06	0.016
Maximum Width	Clast	Quartz	10.59	0.001
Maximum Thickness	Clast	Quartz	8.85	<0.001

Table 8.12 Results of least significance difference test (LSD) for cores on different raw materials in the KZ2 assemblage. Only those results that returned a significant difference are displayed.

Variable	Material	Material	Mean difference	*p*
Maximum Length	Clast	Non-clast	19.58	<0.001
	Clast	Quartz	9.43	<0.001
Maximum Width	Clast	Non-clast	13.66	<0.001
	Clast	Quartz	7.22	<0.001
Maximum Thickness	Clast	Non-clast	10.59	<0.001
	Clast	Quartz	4.25	<0.001

CN1 and CN3 have similar results for core size between the three raw materials. A pattern for core size is harder to discern in the KZ1 assemblage. Non-clast silcrete cores have the largest average dimensions at KZ1, but this is only significantly different to those from KZ2 (Table 8.14). Additionally, clast silcrete cores have the second largest mean dimensions, but this does not hold true for quartz at KZ1, where the average dimensions are the second smallest behind quartz cores in the KZ2 assemblage. The results of core size at KZ1 can be compared to the analysis of nodule size. The average length of cortical flakes was the greatest of all the assemblages at KZ1 for both clast and non-clast silcrete, but the smallest for quartz. This suggests that the clast and non-clast silcrete nodules transported to KZ1 were on average larger than those transported to the other locations. This difference is only significant for clast maximum thickness with KZ2 and all measures on non-clast silcrete with KZ2. In general, the larger size of clast silcrete and non-clast silcrete nodules and the larger average core size suggest that cores at KZ1 were abandoned at an earlier stage of reduction than cores in the other assemblages. Quartz cores at CN1 and CN3 were larger than KZ1 and KZ2 (Table 8.15). This is consistent with the analysis of nodule size at KZ1.

Summary

Consistent patterns of reduction intensity, particularly for both types of silcrete, are difficult to identify in the technological analyses and analysis of average core size. Quartz cores are the least rotated. The MNF to core ratio suggest that quartz is the least intensively worked material, and the non-cortical to cortical complete flake and the non-cortical to cortical core ratio support this. Results for the two types of silcrete are less clear. In all four assemblages the non-cortical to cortical complete flake and the non-cortical to cortical core ratios are highest for clast silcrete rather than non-

Table 8.13 Results of least significance difference test (LSD) for clast silcrete cores between assemblages. Only those results that returned a significant difference are displayed.

Variable	Assemblage	Assemblage	Mean difference	*p*
Maximum Width	CN1	KZ2	7.55	0.026
Maximum Thickness	KZ1	KZ2	5.48	0.031

Table 8.14 Results of least significance difference test (LSD) for non-clast silcrete cores between assemblages. Only those results that returned a significant difference are displayed.

Variable	Assemblage	Assemblage	Mean difference	*p*
Maximum Length	KZ1	KZ2	15.26	0.014
Maximum Width	KZ1	KZ2	13.54	0.018
Maximum Thickness	KZ1	KZ2	9.19	0.014

Table 8.15 Results of least significance difference test (LSD) for quartz cores between assemblages. Only those results that returned a significant difference are displayed.

Variable	Assemblage	Assemblage	Mean difference	*p*
Maximum Length	CN1	KZ1	5.06	0.006
	CN1	KZ2	5.38	<0.001
	CN3	KZ1	5.73	0.004
	CN3	KZ2	6.05	<0.001
Maximum Width	CN1	KZ1	3.77	0.007
	CN1	KZ2	4.40	<0.001
	CN3	KZ1	3.91	0.010
	CN3	KZ2	4.54	<0.001
Maximum Thickness	CN1	KZ1	2.75	0.008
	CN1	KZ2	3.63	<0.001
	CN3	KZ1	2.72	0.015
	CN3	KZ2	3.60	<0.001

clast silcrete. These results do not follow the MNF to core ratio that generally suggested non-clast cores were the most intensively worked. The MNF to core and the non-cortical to cortical complete flake ratios suggest that core reduction is most intensive in the CN1 and KZ2 assemblages, but this is not supported by the non-cortical to cortical core ratio.

In general, the non-cortical to cortical ratios conflict with the MNF to core ratio and, therefore it is difficult to draw straightforward conclusions about the intensity of core reduction. Instead, the variability hints at the complex nature of assemblage formation, and suggests that the Pine Point-Langwell assemblages do not represent a single process of raw material acquisition and reduction. Rather, it seems that the assemblages represent multiple processes, which may include the recycling of cores and flakes. Certainly, this would help explain some of the inconsistencies between the measures of core reduction intensity. Analysis of certain flake characteristics (platform type, flake size and shape) may provide a clearer picture of the relative intensity of raw material reduction both between raw materials and assemblages.

At least some aspects of variability between the assemblages reflect differential access to raw materials. For example, the non-intensive reduction of quartz reflects the abundant sources of this material within the immediate vicinity of the assemblages. Measures of silcrete reduction are full of inconsistencies; while there are some aspects of a distance decay relationship, assemblages located furthest from possible silcrete sources do not demonstrate a clear pattern of more intensive reduction. This may suggest a number of things; silcrete was not always transported to the locations from the closest sources, each location has variable occupation histories, and a single place was rarely used the same way through time.

Flake Production

The shape, size and physical characteristics of flakes provide information on core reduction strategies. In turn, these relate to the overall intensity of raw material reduction and assemblage composition. Therefore, an analysis of flake production is important to understanding the temporal character of assemblage formation.

Complete Flake Exterior Platform Characteristics

The character of the exterior platform on the dorsal surface of a flake provides information about the surface of the core from which the flake was struck (Holdaway and Stern 2004), including the degree to which core platforms were prepared prior to flake removal. The main type of core platform preparation involves the removal of the overhang left on the core surface from previous flake removals. This may inhibit

Table 8.16 Frequency and percentage (in brackets) of exterior platform surface of complete flakes per raw material for the CN1 assemblage. Flake scars are the most common exterior platform surface for each raw material. Trimming is most common on non-clast silcrete, followed by clast.

Exterior Platform Surface	Clast	Non-clast	Quartz
Cortex	11 (2.4)	3 (2)	544 (20.1)
Flake Scar	415 (90.2)	118 (77.6)	2115 (78.2)
Trimming	34 (7.4)	31 (20.4)	47 (1.7)
Total	460 (100)	152 (100)	2706 (100)

Table 8.17 Frequency and percentage (in brackets) of exterior platform surface of complete flakes per raw material for the CN3 assemblage. Flake scars are the most common exterior platform surface for each raw material. Trimming is most common on non-clast silcrete, followed by clast.

Exterior Platform Surface	Clast	Non-clast	Quartz
Cortex	7 (2.5)	5 (6.8)	342 (20.2)
Flake Scar	256 (91.8)	54 (74)	1337 (79.2)
Trimming	16 (5.7)	14 (19.2)	10 (0.6)
Total	279 (100)	73 (100)	1689 (100)

Table 8.18 Frequency and percentage (in brackets) of exterior platform surface of complete flakes per raw material for the KZ1 assemblage. Flake scars are the most common exterior platform surface for each raw material. Trimming is most common on non-clast silcrete, followed by clast.

Exterior Platform Surface	Clast	Non-clast	Quartz
Cortex	9 (2.3)	5 (5.2)	77 (11)
Flake Scar	333 (85.4)	77 (79.4)	600 (85.8)
Trimming	48 (12.3)	15 (15.5)	22 (3.1)
Total	390 (100)	97 (100)	699 (100)

Table 8.19 Frequency and percentage (in brackets) of exterior platform surface of complete flakes per raw material for the KZ2 assemblage. Flake scars are the most common exterior platform surface for each raw material. Trimming is most common on non-clast silcrete, followed by clast.

Exterior Platform Surface	Clast	Non-clast	Quartz
Cortex	44 (3)	9 (4.2)	299 (12.3)
Flake Scar	1399 (93.8)	187 (87.4)	2106 (86.9)
Trimming	48 (3.2)	18 (8.4)	19 (0.8)
Total	1491 (100)	214 (100)	2424 (100)

the removal of subsequent flakes and therefore its removal, which leaves a series of small flake scars on the core surface, demonstrates an effort to minimise the chances of producing undesirable flake forms. Core preparation can be expected in assemblages where there is an incentive to efficiently manage raw material stocks (Holdaway and Stern 2004).

Three types of exterior platform characteristic were recorded: cortex, scar and trimming. The proportion of different exterior platform characteristics is presented in Tables 8.16 to 8.19.

Cortex on the exterior platform surface is uncommon on both clast and non-clast silcrete flakes. The highest proportion for clast silcrete occurs at KZ2 (3%), while for non-clast silcrete it is at CN3 (6.8%). Cortex is far more common on the exterior platform of quartz complete flakes and accounts for between 20.2% at CN3 and 11% at KZ1. This supports the conclusion that quartz nodules contained a large proportion of cortex whereas clast and non-clast silcrete nodules were more likely to be partially decortified before being transported to the locations.

Flake scar refers to negative flake scars initiated from the proximal end of the dorsal surface of a flake. These lack the trimming indicative of core preparation. Flake scars are the most common exterior platform surface, but the proportions vary between materials and assemblages. The proportion of flake scar is highest for clast silcrete in the CN1, CN3 and KZ2 assemblages. At KZ1 the proportion for clast silcrete is slightly less than that for quartz. The proportion of flake scar is high for quartz at KZ1, where the proportion of cortex is the lowest of the four assemblages. This suggests that either cortical quartz nodules were less often transported to KZ1 or that the reduction of quartz nodules was more intensive than in the other assemblages. This has the effect of increasing the proportion of flakes with dorsal flake scars relative to those with cortex. Analysis of core reduction in the previous section indicated that quartz cores at KZ1 were not significantly more reduced than cores in the other assemblages. Therefore, it can be concluded that non-cortical nodules were more often introduced to KZ1 than at the other locations. Non-clast silcrete has the lowest proportion of flake scar in the CN1, CN3 and KZ1 assemblages, but at KZ2 the proportion is slightly larger than that for quartz.

Trimming on the exterior platform surface is indicative of core preparation. In each of the assemblages the proportion of trimming is highest for non-clast silcrete, varying between 20.4% at CN1 and 8.4% at KZ2. Clast silcrete has the next largest proportion of trimming, varying between 12.3% at KZ1 and 3.2% at KZ2. Trimming is rare on quartz, varying between 3.1% at KZ1 and 0.6% at CN3. The fact that trimming is highest on non-clast silcrete in each of the assemblages is consistent with increasing the efficiency of reducing high quality non-local material through overhang removal from the platform surface of cores.

The KZ2 assemblage demonstrates proportionally less core platform preparation for clast and non-clast silcrete than the other three assemblages. In addition, the value for quartz trimming is the second lowest of the four assemblages. Not surprisingly, KZ2 has the highest value for flake scar in each of the raw materials. This indicates that core platform preparation was not as regular a feature of flake production techniques at KZ2 compared to the other assemblages.

Flake Size

As core reduction proceeds, the size of both cores and flakes/manufacturing debris decreases because stone working is a reductive technology (Holdaway and Stern 2004). Flakes will on average be smaller in assemblages where core reduction is intensive.

The mean length, width, thickness, platform width and platform thickness of complete flakes by raw material and assemblage are presented in Tables 8.20 to 8.22. Clast silcrete complete flakes in all of the assemblages have the widest platforms and are on average longer and wider than complete flakes manufactured from either quartz or non-clast silcrete. These differences are significant. Quartz complete flakes are on average thicker and have thicker platforms than complete flakes in both the silcrete types. Non-clast silcrete flakes are on average smaller in every dimension than those of clast silcrete and quartz with the exception of length in the KZ1 assemblage where the mean length of non-clast complete flakes is .1 millimetres greater than that of the mean length for quartz. The results for complete flake size largely follow those for the mean length of complete cortical flakes, which indicated that clast silcrete nodules were the largest followed by quartz and non-clast silcrete. Significant results between materials within individual assemblages are presented in Tables 8.23 to 8.26.

Clast silcrete flakes in the KZ1 assemblage are on average larger in every dimension except platform thickness, which is greater at CN1. Table 8.27 indicates a significant difference between clast silcrete complete flakes in the KZ1 assemblage and the other three assemblages, with the exception of CN1 for width and both the platform measurements. These results are similar to tests based on cortical complete flake length, which indicated that clast silcrete flakes are longest at KZ1. KZ1 and KZ2 have significant differences between each other and the two CN assemblages. The difference between CN1 and CN3 assemblages is only significant for both platform measurements. The difference in each of the remaining mean values between CN1 and CN3 is not significant.

For non-clast silcrete, complete flakes in the KZ1 assemblage are on average larger than complete flakes in the other three assemblages (Table 8.28). This is consistent with the large average length of complete cortical flakes at KZ1. The mean values for length, width, thickness and

Table 8.20 Mean length, width, thickness, platform width and platform thickness of clast silcrete complete flakes in the four assemblages. length F=15.14, *df* 3, 2632, p=<0.001, width F=21.02, *df* 3, 2632, p=<0.001, thickness F=22.42, *df* 3, 2632, p=<0.001, platform width F=11.51, *df* 3, 2376, p=<0.001, platform thickness F=13.57, *df* 3,2376, p=<0.001

Assemblage		Length (mm)	Width (mm)	Thickness (mm)	Platform Width (mm)	Platform Thickness (mm)
CN1	Mean	26.8	22.8	7.8	17.7	6.9
	S.D.	11.2	9.2	4.5	9.1	4.2
	N	466	466	466	425	425
CN3	Mean	26.4	22	7.6	16.2	6.2
	S.D.	12	9.8	4.9	8.8	4.1
	N	280	280	280	256	256
KZ1	Mean	28.5	23.7	8.4	17.9	6.9
	S.D.	11	8.8	4.7	8.7	3.9
	N	398	398	398	332	332
KZ2	Mean	24.9	20.3	6.6	15.6	5.8
	S.D.	9.4	8.6	3.9	8.1	3.7
	N	1492	1492	1492	1367	1367

Table 8.21 Mean length, width, thickness, platform width and platform thickness of non-clast silcrete complete flakes in the four assemblages. length F=6.07, *df* 3, 536, p=<0.001, width F=5.21, *df* 3, 535, p=<0.001, thickness F=4.28, *df* 3, 536, p=0.005 platform width F=3.43, *df* 3, 469, p=0.017, platform thickness F=2.08, *df* 3, 470, p=0.103)

Assemblage		Length (mm)	Width (mm)	Thickness (mm)	Platform Width (mm)	Platform Thickness (mm)
CN1	Mean	21.1	17.9	5.8	13	4.6
	S.D.	5.6	6	2.9	6.1	2.7
	N	153	153	153	137	137
CN3	Mean	21.4	18.1	5.7	11.8	4.3
	S.D.	6	6	2.9	5.4	2.1
	N	73	73	73	66	66
KZ1	Mean	24.5	21	7.1	15	5.4
	S.D.	7.5	7.9	3.5	6.5	2.9
	N	100	99	100	82	83
KZ2	Mean	21.8	18.2	6.1	13.1	4.8
	S.D.	6.8	6.5	2.8	6.4	2.8
	N	214	214	214	188	188

Table 8.22 Mean length, width, thickness, platform width and platform thickness of quartz complete flakes in the four assemblages (complete bipolar flakes not included). length, F=11.78 *df* 3, 7540, p=<0.001, width, F =24.83, *df* 3, 7540, p=<0.001, thickness, F=42.81, *df* 3, 7540, p =<0.001, platform width F=29.37, *df* 3, 6656, p=<0.001, platform thickness F=41.25, *df* 3, 6656, p=<0.001

Assemblage		Length (mm)	Width (mm)	Thickness (mm)	Platform Width (mm)	Platform Thickness (mm)
CN1	Mean	25.3	20.9	8.6	16.5	7
	S.D.	7.9	6.9	4	7.1	3.5
	N	2719	2719	2719	2402	2402
CN3	Mean	25.5	20.7	8.4	16	6.9
	S.D.	7.9	7.5	4.3	7.3	4
	N	1689	1689	1689	1493	1493
KZ1	Mean	24.4	20.9	8.6	16.7	7.5
	S.D.	8	6.9	4.2	7.2	5.2
	N	712	712	712	582	582
KZ2	Mean	24.3	19.3	7.4	14.7	6
	S.D.	7	6.7	3.5	6.8	3.1
	N	2424	2424	2424	2183	2183

Table 8.23 Results of least significance difference test (LSD) for complete flakes on different raw materials in the CN1 assemblage. Only those results that returned a significant difference are displayed.

Variable	Material	Material	Mean difference	p
Length	Clast	Non-clast	5.73	<0.001
	Clast	Quartz	1.59	<0.001
	Non-clast	Quartz	-4.14	<0.001
Width	Clast	Non-clast	4.95	<0.001
	Clast	Quartz	2.0	<0.001
	Non-clast	Quartz	-2.95	<0.001
Thickness	Clast	Non-clast	1.94	<0.001
	Clast	Quartz	-0.84	<0.001
	Non-clast	Quartz	-2.79	<0.001
Platform Width	Clast	Non-clast	4.78	<0.001
	Clast	Quartz	1.21	<0.001
	Non-clast	Quartz	-3.57	<0.001
Platform Thickness	Clast	Non-clast	2.28	<0.001
	Non-clast	Quartz	-2.39	<0.001

Table 8.24 Results of least significance difference test (LSD) for complete flakes on different raw materials in the CN3 assemblage. Only those results that returned a significant difference are displayed.

Variable	Material	Material	Mean difference	p
Length	Clast	Non-clast	4.94	<0.001
	Non-clast	Quartz	-4.04	<0.001
Width	Clast	Non-clast	3.86	<0.001
	Clast	Quartz	1.30	0.010
	Non-clast	Quartz	-2.56	0.006
Thickness	Clast	Non-clast	1.84	0.001
	Clast	Quartz	-0.83	0.003
	Non-clast	Quartz	-2.67	<0.001
Platform Width	Clast	Non-clast	4.35	<0.001
	Non-clast	Quartz	-4.19	<0.001
Platform Thickness	Clast	Non-clast	1.89	0.001
	Clast	Quartz	-0.71	0.008
	Non-clast	Quartz	-2.60	<0.001

Table 8.25 Results of least significance difference test (LSD) for complete flakes on different raw materials in the KZ1 assemblage. Only those results that returned a significant difference are displayed.

Variable	Material	Material	Mean difference	*p*
Length	Clast	Non-clast	4.05	<0.001
	Clast	Quartz	4.10	<0.001
Width	Clast	Non-clast	2.73	0.002
	Clast	Quartz	2.85	<0.001
Thickness	Clast	Non-clast	1.30	0.007
	Non-clast	Quartz	-1.58	0.001
Platform Width	Clast	Non-clast	2.90	0.002
	Clast	Quartz	1.20	0.023
Platform Thickness	Clast	Non-clast	1.54	0.007
	Non-clast	Quartz	-2.09	<0.001

Table 8.26 Results of least significance difference test (LSD) for complete flakes on different raw materials in the KZ2 assemblage. Only those results that returned a significant difference are displayed.

Variable	Material	Material	Mean difference	*p*
Length	Clast	Non-clast	3.10	<0.001
	Clast	Quartz	.62	0.018
	Non-clast	Quartz	-2.47	0.018
Width	Clast	Non-clast	2.08	<0.001
	Clast	Quartz	.99	<0.001
	Non-clast	Quartz	-1.08	0.41
Thickness	Clast	Non-clast	.53	0.047
	Clast	Quartz	-.83	<0.001
	Non-clast	Quartz	-1.36	<0.001
Platform Width	Clast	Non-clast	2.46	<0.001
	Clast	Quartz	.84	<0.001
	Non-clast	Quartz	-1.63	0.003
Platform Thickness	Clast	Non-clast	.97	<0.001
	Clast	Quartz	-.22	0.054
	Non-clast	Quartz	-1.19	<0.001

Table 8.27 Results of least significance difference test (LSD) for clast silcrete complete flakes between assemblages. Only those results that returned a significant difference are displayed.

Variable	Assemblage	Assemblage	Mean difference	*p*
Length	CN1	KZ1	-1.69	0.016
	CN1	KZ2	1.97	<0.001
	CN3	KZ1	-2.16	0.007
	CN3	KZ2	1.50	0.025
	KZ1	KZ2	3.66	<0.001
Width	CN1	KZ2	2.55	<0.001
	CN3	KZ1	-1.76	0.011
	CN3	KZ2	1.64	0.005
	KZ1	KZ2	3.39	<0.001
Thickness	CN1	KZ1	-0.59	0.041
	CN1	KZ2	1.15	<0.001
	CN3	KZ1	-0.81	0.015
	CN3	KZ2	0.93	0.001
	KZ1	KZ2	1.74	<0.001
Platform Width	CN1	CN3	1.55	0.021
	CN1	KZ2	2.19	<0.001
	CN3	KZ1	-1.70	0.016
	KZ1	KZ2	2.33	<0.001
Platform Thickness	CN1	CN3	0.73	0.016
	CN1	KZ2	1.13	<0.001
	CN3	KZ1	-0.71	0.027
	KZ1	KZ2	1.10	<0.001

Table 8.28 Results of least significance difference test (LSD) for non-clast silcrete complete flakes between assemblages. Only those results that returned a significant difference are displayed.

Variable	Assemblage	Assemblage	Mean difference	*p*
Length	CN1	KZ1	-3.38	<0.001
	CN3	KZ1	-3.05	0.002
	KZ2	KZ1	2.71	0.001
Width	CN1	KZ1	-3.06	<0.001
	CN3	KZ1	-2.89	0.005
	KZ2	KZ1	-2.75	0.001
Thickness	CN1	KZ1	-1.24	0.001
	CN3	KZ1	-1.35	0.003
	KZ2	KZ1	-0.97	0.007
Platform Width	CN1	KZ1	-2.02	0.020
	CN3	KZ1	-3.15	0.002
	KZ2	KZ1	-1.90	0.021
Platform Thickness	CN3	KZ1	-1.06	0.018

Table 8.29 Results of least significance difference test (LSD) for quartz complete flakes between assemblages. Only those results that returned a significant difference are displayed.

Variable	Assemblage	Assemblage	Mean difference	p
Length	CN1	KZ1	0.82	0.011
	CN1	KZ2	1.0	<0.001
	CN3	KZ1	1.04	0.002
	CN3	KZ2	1.22	<0.001
Width	CN1	KZ2	1.55	<0.001
	CN3	KZ2	1.33	<0.001
	KZ1	KZ2	1.54	<0.001
Thickness	CN1	KZ2	1.16	<0.001
	CN3	KZ2	0.93	<0.001
	KZ1	KZ2	1.19	<0.001
Platform Width	CN1	CN3	0.51	0.028
	CN1	KZ2	1.82	<0.001
	CN3	KZ2	1.31	<0.001
	KZ1	KZ2	1.97	<0.001
Platform Thickness	CN1	KZ1	-0.42	0.015
	CN1	KZ2	1.02	<0.001
	CN3	KZ1	-0.56	0.002
	CN3	KZ2	0.88	<0.001
	KZ1	KZ2	1.44	<0.001

platform width are significantly different between KZ1 and the other three assemblages, but for platform thickness the difference is only significant between KZ1 and CN3.

The average dimensions of quartz complete flakes are larger in both the CN1 and CN3 assemblages when compared to the KZ1 and KZ2 assemblages (Table 8.29). Some of these differences, particularly length, are significantly different. This may reflect the larger size of quartz nodules at CN1 and CN3 compared to KZ1 and KZ2. There are other differences in quartz flake size between the assemblages, but significant differences are not consistent.

The results of the analysis of complete flake size indicate that flakes are on average larger in the KZ1 assemblage compared to the three other assemblages. This result largely follows that of the analysis of nodule size in Chapter Seven, which indicated that the KZ1 assemblage had the largest clast silcrete and non-clast silcrete nodules. The same is true for quartz complete flake size, where the average size of flakes in the two CN assemblages is larger than that of the KZ1 and KZ2 assemblages. This result is consistent with the analysis of nodule size, which indicated that quartz nodules are on average larger in the CN1 and CN3 assemblages than those in the KZ1 and KZ2 assemblages.

Flake Shape

Maximising the surface area of a flake relative to thickness increases the amount of edge available for tool resharpening (Pelcin 1997, Roth and Dibble 1998). Roth and Dibble (1998) argue that this is particularly desirable if flakes rather than cores are transported. If all stages of core reduction were represented in the Pine Point-Langwell assemblages, flakes with a wide range of different surface areas would be expected. Flake shape can be presented on a line graph, which plots platform width/platform thickness on the x-axis against the average of surface area (flake length times flake width) divided by flake thickness on the y-axis (Dibble 1997).

Flake shape for each raw material type is plotted in Figures 8.1 to 8.3. There is a general linear trend of increasing complete flake surface area relative to thickness as the platform width to platform thickness ratio increases.

For a given platform width to platform thickness ratio clast silcrete complete flakes have a larger surface area relative to thickness than complete flakes made on non-clast silcrete and quartz. Non-clast silcrete shows a variable pattern that may reflect the small number of cases for certain platform width to platform thickness ratios. Quartz exhibits a much more uniform linear pattern of increasing surface area relative to thickness with increased platform width relative to platform thickness.

Patterns of inter assemblage variability within the same raw material class are evident, especially for clast and non-clast silcrete. Complete clast silcrete flakes at CN3 generally have on average larger surface areas relative to thickness than clast flakes in other assemblages, with the exception of a platform width to platform thickness ratio of three and 3.5. KZ2 complete clast silcrete flakes generally have the lowest value, followed by CN1 and KZ1. Non-clast silcrete complete flakes in all of the assemblages demonstrate a variable pattern that is difficult to interpret. As suggested above, this reflects the small number of complete flakes of this material and any interpretation must be accepted with caution. KZ2 complete flakes generally have the lowest values followed by CN1, while KZ1 have the highest values until a platform width to platform thickness ratio greater than four is reached. The plot for quartz indicates a tight pattern and exhibits the least amount of variability between assemblages. KZ1 complete flakes generally have the lowest values, while CN3 and KZ2 complete flakes have similar values until a platform width to platform thickness ratio of three at which point CN3 values are greater. CN1 has the lowest values for a platform width to platform thickness ratio less than four.

Clast silcrete complete flakes have the greatest amount of surface area available for tool manufacture and quartz the lowest. Clast silcrete flakes have a longer potential edge to either resharpen or use, than flakes of either non-clast silcrete or quartz. Therefore, clast silcrete flakes may have been preferred for tool production compared to quartz flakes. Understanding how non-clast flake shape influenced tool manufacture requires consideration of the tool forms produced on this material.

Flake Platform Type

The characteristics of flake platforms reflect the nature of the core platform from which they were removed. Different platform types provide not only information about the type of nodules reduced, but also about the different techniques applied to core reduction. All flakes that retain information on platform type are included in the following analysis. The proportion of different platform types per raw material is presented in Tables 8.30 to 8.32.

Uni platforms are the most common platform type across the three raw materials and the four assemblages, but there are considerable differences in the proportion of uni platforms for both types of silcrete compared to quartz. Uni platforms account for 80% to 85% of platform types in clast and non-clast silcrete compared to 65% to 75% for quartz. Cortical platforms are barely represented in

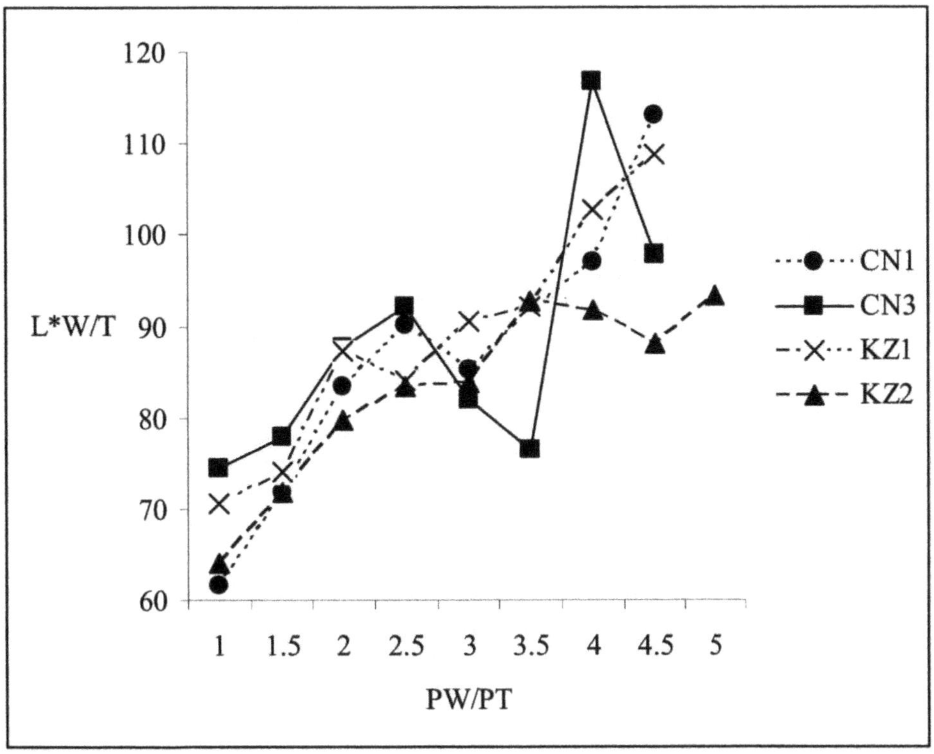

Figure 8.1 Shape plot of clast complete flakes by assemblage. This indicates that clast flakes from the CN3 assemblage on average have a larger surface area to thickness ratio clast silcrete flakes from the other assemblages.

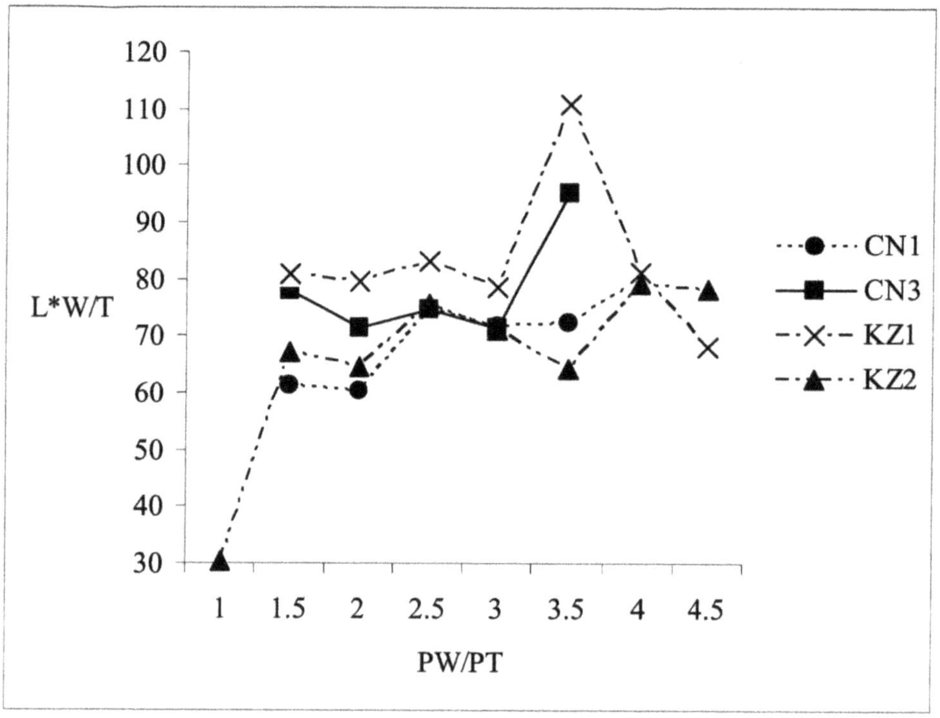

Figure 8.2 Shape plot of non-clast complete flakes by assemblage. The small number of flakes in some assemblages creates a variable pattern. However, the plot indicates that non-clast flakes from the KZ1 assemblage on average have a larger surface area to thickness ratio than non-clast flakes from the other assemblages.

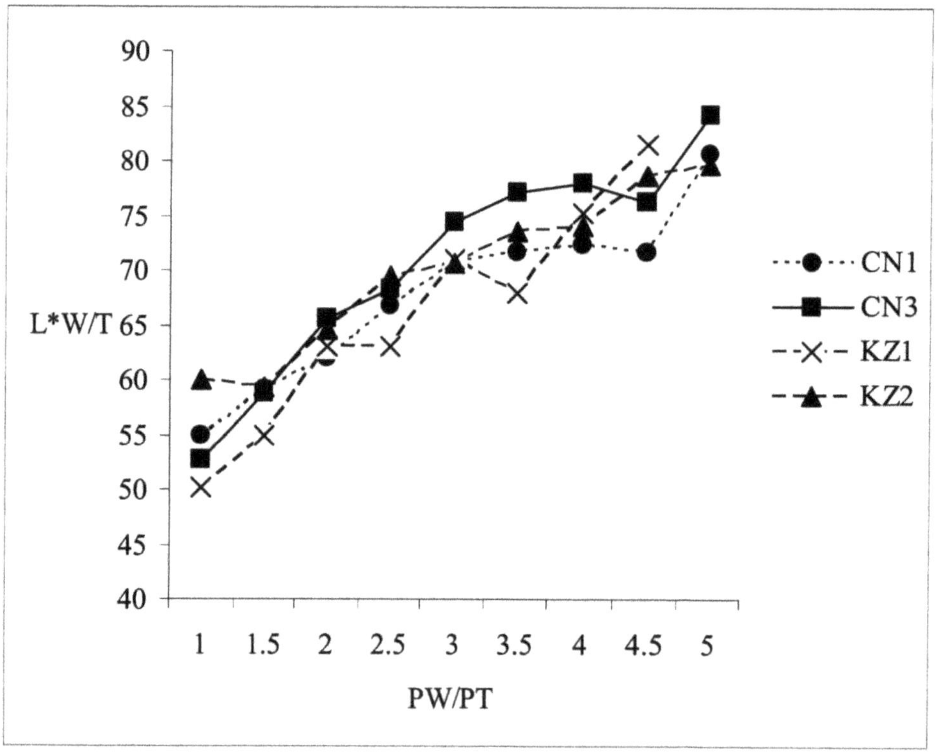

Figure 8.3 Shape plot of quartz complete flakes by assemblage. The plot indicates that quartz flakes from the CN3 assemblage on average have a larger surface area to thickness ratio than quartz flakes from the other assemblages. However, flakes in the four assemblages are very similar in shape and it is difficult to clearly separate them.

either type of silcrete, however they make up between 15% and 25% of platform types for quartz. Bi platforms are uncommon amongst the three raw material categories, accounting for approximately 10% of platform types on clast and non-clast silcrete and 5% of quartz platform types in all assemblages. Crushed platforms are common only in quartz, however the proportion of these is generally low (5%-7%) across the four assemblages.

Bi platforms represent a core surface that has been struck from at least two different directions and are therefore indicative of core rotation. With the largest proportion of rotated core forms in the KZ1 assemblage, this assemblage may be expected to contain the highest proportion of bi platforms. However, comparisons per raw material between the assemblages indicate that the KZ2 assemblage rather the KZ1 assemblage has the largest proportion of bi platforms on each of the three raw materials, although the differences between KZ2 and the other assemblages are only slight. For clast silcrete the KZ2 bi platform proportion of 12% is only marginally larger than that for CN1 with 9.6% while the same is true for non-clast silcrete where the proportion at KZ2 is 14.2% compared to 13.3% at CN3. The proportion for quartz is similar between the assemblages. This may suggest that core rotation occurred late in the reduction sequence at KZ1, perhaps immediately prior to discard and therefore few flakes were removed following core rotation.

Cortical platforms indicate that the core platform from which the flake was struck contained cortex and is likely to be representative of the earlier stages of core reduction

Table 8.30 Clast silcrete platform type frequency and percentage (in brackets) for each assemblage. Uni platforms are the most common type across all assemblages, followed by bi platforms. Cortical and crushed platforms are uncommon.

Platform type	CN1	CN3	KZ1	KZ2
Bi	50 (9.6)	25 (8.2)	39 (9.4)	209 (12)
Cortical	12 (2.3)	3 (1)	7 (1.7)	26 (1.5)
Crushed	1 (0.2)		8 (1.9)	15 (0.9)
Uni	458 (87.9)	276 (90.8)	361 (87)	1496 (85.7)
Total	521 (100)	304 (100)	415 (100)	1746 (100)

Table 8.31 Non-clast silcrete platform type frequency and percentage (in brackets) for each assemblage. Uni platforms are the most common type across all assemblages, followed by bi platforms, although the proportion of these is greater than that for clast silcrete. Cortical and crushed platforms are uncommon.

Platform type	CN1	CN3	KZ1	KZ2
Bi	19 (9.6)	12 (13.3)	14 (11.4)	39 (14.2)
Cortical	4 (2)	1 (1.1)	4 (3.3)	7 (2.6)
Crushed			1 (0.8)	3 (1.1)
Uni	174 (88.3)	77 (85.6)	104 (84.6)	225 (82.1)
Total	197 (100)	90 (100)	123 (100)	274 (100)

Table 8.32 Quartz platform type frequency and percentage (in brackets) for each assemblage. Uni platforms are the most common type for quartz, however the proportions of cortical platforms are greater than that for the other raw material types. This supports earlier conclusions that cortical quartz nodules were predominantly reduced at the locations.

Platform type	CN1	CN3	KZ1	KZ2
Bi	140 (2.8)	85 (4.9)	36 (4.4)	168 (6.2)
Cortical	765 (26.1)	438 (25.1)	129 (15.7)	411 (15.1)
Crushed	124 (4.2)	47 (2.7)	74 (9)	106 (3.9)
Uni	1907 (65)	1176 (67.4)	582 (70.9)	2045 (74.9)
Total	2936 (100)	1746 (100)	821 (100)	2730 (100)

Table 8.33 Flake to MNT (minimum number of tools) ratio per raw material and assemblage. Proportionally more non-clast silcrete flakes are retouched into tools, followed by clast silcrete in three assemblages (except KZ1 where the quartz ratio is lower). Quartz flakes are generally the least frequently resharpened.

Material	CN1	CN3	KZ1	KZ2
Clast	8.3	10.5	11.7	13
Non-clast	3.8	2.8	4.3	3.9
Quartz	13.9	18	9.5	19.9

Table 8.34 Complete tool form frequency and percentage (in brackets) per raw material type in the CN1 assemblage. Scrapers are the most common tool form within each raw material class. There are less notched artefacts (denticulates and notches) on non-clast silcrete and more utilised artefacts compared to clast silcrete and quartz.

Tool type	Clast	Non-clast	Quartz
Backed blade	1 (1.5)		10 (4.3)
Denticulate	9 (13.8)	2 (4.3)	15 (6.4)
Notch	5 (7.7)	3 (6.5)	43 (18.5)
Scraper	46 (70.8)	19 (41.3)	149 (63.9)
Thumbnail Scraper	1 (1.5)	1 (2.2)	1 (0.4)
Tula slug		7 (15.2)	
Utilised	3 (4.6)	14 (30.4)	15 (6.4)
Total	65 (100)	46 (100)	233 (100)

Table 8.35 Complete tool form frequency and percentage (in brackets) per raw material type in the CN3 assemblage. Scrapers are the most common tool form within each raw material class. There are less notched artefacts (denticulates and notches) on non-clast silcrete and more utilised artefacts compared to clast silcrete and quartz.

Tool type	Clast	Non-clast	Quartz
Backed blade	1 (3.3)		8 (7.3)
Burin		1 (4.3)	
Burren slug		1 (4.3)	
Denticulate	2 (6.7)		9 (8.3)
Notch	2 (6.7)	1 (4.3)	17 (15.6)
Pirri	3 (10)	1 (4.3)	1 (0.9)
Scraper	19 (63.3)	11 (47.8)	64 (58.7)
Thumbnail scraper		2 (8.7)	
Tula slug	2 (6.7)	3 (13)	
Utilised	1 (3.3)	3 (13)	10 (9.2)
Total	30 (100)	23 (100)	109 (100)

(Baulmer 1988). The proportion of cortical platforms on clast and non-clast silcrete is very low across all assemblages, a pattern that is expected given the likelihood that partially decortified nodules were transported to, and reduced at, each location. The opposite is true of quartz, where approximately 25% of platforms at CN1 and CN3 are cortical with this decreasing to about 15% at KZ1 and KZ2. This supports the results of the analysis of nodule form, which suggested that cortical quartz nodules were frequently transported to and reduced at the four locations.

Summary

The analysis of flake attributes has revealed a number of patterns in flake production between raw materials and assemblages, with these demonstrating greater consistencies than the characteristics of core reduction. Clast silcrete complete flakes are on average larger than quartz and non-clast silcrete complete flakes, which are the smallest. This is consistent with the analysis of nodule size, which indicated clast silcrete nodules were larger than non-clast silcrete and quartz nodules. KZ1 has the largest average complete flake size for clast silcrete and non-clast silcrete, however for quartz both the CN assemblages have the largest flakes. There is little core platform preparation, the majority of which occurs on non-clast silcrete flakes followed by clast silcrete. Larger proportions of quartz complete flakes compared to those of either type of silcrete have cortex on the exterior platform, again indicative of the exploitation of local material. The CN1 assemblage has the most evidence for platform preparation for non-clast silcrete and KZ1 for clast silcrete. The large proportion of uni platforms confirms the earlier analysis of core form that indicated core rotation was not as common as unifacial reduction. Bi platforms are most common in non-clast silcrete, indicating that core rotation of this material occurred more frequently than in other raw material types. Clast silcrete complete flakes have a larger surface area relative to thickness for a given platform width/platform thickness ratio than is true for other raw materials. Complete clast flakes at CN3 on average have the greatest surface area relative to thickness. Non-clast silcrete shows a variable pattern arising from a limited number of individual cases. Quartz demonstrates the least variability between assemblages.

In conclusion, the analysis of flake characteristics indicates some general technological trends in flake production between the assemblages. These are largely consistent with other measures of nodule and core size. Combined, the analysis of flake production suggests a degree of consistency in flake production techniques. This consistency contrasts to the variability evident in the measures of raw material reduction intensity.

Tool Production, Resharpening and Discard

Retouched flakes are an important component of stone artefact assemblages. As discussed in Chapter Three, the proportional representation of different tool types may reflect the temporal character of past occupations. At individual locations tool manufacture, resharpening and discard are linked to the frequency and duration of occupation. However, these factors should not be assessed independent of core reduction and flake manufacture.

Flake to Minimum Number of Tools Ratio

The flake to MNT ratio is the simplest measure of tool production. Low values for this ratio indicate that proportionally more flakes in an assemblage are modified into tools. Proportionally fewer quartz flakes are modified into tools in the CN1, CN3 and KZ2 assemblages compared to both types of silcrete (Table 8.33). The KZ1 assemblage is an exception. Here the ratio for quartz is less than that for clast silcrete. The ratio in all assemblages is lowest for non-clast silcrete indicating that there are fewer unretouched flakes relative to retouched flakes on this material than in any other material. Non-clast silcrete appears to have been favoured for tool production.

In general, tool production is least intensive at KZ2. The ratio for clast silcrete and quartz at KZ2 is the highest of the assemblages; the value for non-clast silcrete is the second highest behind KZ1. The CN1 and CN3 assemblages have the lowest values for both types of silcrete, however there is only a marginal difference between the non-clast value at CN1 and KZ2. Additionally, there is no clear pattern between the silcretes in the CN1 and CN3 assemblages, with the value for clast lowest at CN1 and the value for non-clast silcrete lowest at CN3. For quartz, the values in both CN1 and CN3 is much higher than that at KZ1, with CN3 having the second highest value across all four of the assemblages.

Complete Tool Form Proportions

The frequency and proportion of different tool forms in the four assemblages are presented in Tables 8.34 to 8.37. CN3 has the greatest number of tool forms with a total of ten, nine at KZ2, and seven each at CN1 and KZ1. There is no clear relationship between sample size and tool richness. The CN3 assemblage is smaller in terms of the number of individual artefacts than the KZ2 and CN1 assemblages, yet it contains the greatest number of tool forms.

Backed blades, burren slugs, burins, tula slugs, pirri points, and thumbnail scrapers are generally rare. Tula slugs are only manufactured from clast silcrete and more commonly non-clast silcrete. This pattern has been identified in other areas of western New South Wales (Holdaway *et al.* 2004). Denticulates and single notches are slightly more common along with utilised flakes. Scrapers are by far the most common tool form for each raw material type and assemblage, a pattern that is also consistent with other western New South Wales assemblages (e.g., Holdaway *et al.* 2004, Shiner *et al.* 2005). Clast silcrete has the largest proportion of scrapers in the CN1, CN3 and KZ1 assemblages. At KZ2 the proportion of scrapers is greatest for non-clast silcrete. The proportion of scrapers is lowest for non-clast silcrete at both CN1 and

Table 8.36 Complete tool form frequency and percentage (in brackets) per raw material type in the KZ1 assemblage. Scrapers are the most common tool form within each raw material class. There are less notched artefacts (denticulates and notches) on non-clast silcrete and more utilised artefacts compared to clast silcrete and quartz.

Tool type	Clast	Non-clast	Quartz
Backed blade		1 (3.6)	2 (2.2)
Burren slug	1 (2.6)	1 (3.6)	
Denticulate	6 (15.8)	1 (3.6)	15 (16.9)
Notch	6 (15.8)	1 (3.6)	16 (18)
Scraper	25 (65.8)	18 (64.3)	54 (60.7)
Tula slug		1 (3.6)	
Utilised		5 (17.9)	2 (2.2)
Total	38 (100)	28 (100)	89 (100)

Table 8.37 Complete tool form frequency and percentage (in brackets) per raw material type in the KZ2 assemblage. Scrapers are the most common tool form within each raw material class. There are less notched artefacts (denticulates and notches) on non-clast silcrete and more utilised artefacts compared to clast silcrete and quartz.

Tool type	Clast	Non-clast	Quartz
Backed blade	12 (8.7)	1 (1.6)	17 (11.6)
Burren slug	1 (0.7)	1 (1.6)	
Denticulate	3 (2.2)	1 (1.6)	18 (12.2)
Notch	14 (10.1)	1 (1.6)	12 (8.2)
Pirri	6 (4.3)		4 (2.7)
Scraper	78 (56.5)	36 (58.1)	83 (56.5)
Thumbnail scraper	2 (1.4)	4 (6.5)	4 (2.7)
Tula slug	1 (0.7)	6 (9.7)	
Utilised	21 (15.2)	12 (19.4)	9 (6.1)
Total	138 (100)	62 (100)	147 (100)

CN3, but at KZ1 there are proportionally more non-clast silcrete scrapers than for quartz. The frequency of scrapers can be ranked per raw material and assemblage (Table 8.38). Generally CN1 has a large proportion of clast and quartz scrapers and KZ2 has low proportions.

Table 8.38 Ranking of the proportion of scrapers per raw material and assemblage. A value of one indicates the highest proportion and a value of four the lowest.

Assemblage	Clast	Non-clast	Quartz
CN1	1	4	1
CN3	3	3	3
KZ1	2	1	2
KZ2	4	2	4

If the proportions of heavily worked tools (denticulates, notches and scrapers) are considered together, there are higher proportions of these tools for clast-silcrete and quartz across all assemblages than is the case for non-clast silcrete. Further, non-clast silcrete has the highest proportion of lightly worked tools (utilised flakes). Non-clast silcrete has the largest proportion of the most intensively worked tool form (tula slug) of the three raw materials, but this is generally not matched by large proportions of scrapers, notches and denticulates in the assemblages.

KZ1 has the highest proportion of intensively worked tool forms (denticulates, notches and scrapers combined) for each raw material class. This is followed by CN1, which has the second highest value for clast silcrete and quartz, but the equal lowest with CN3 for non-clast silcrete. The proportion of intensively worked tool forms is greater at CN3 for clast silcrete and quartz compared to KZ2, but not for non-clast silcrete.

Complete Flake Surface Area to Complete Scraper Surface Area Ratio

The following two analyses focus on scrapers because they are the most common tool form for each raw material in the four assemblages and provide an adequate number of cases for analysis. The KZ1 assemblage contains the highest proportion of intensively worked tools for each raw material class and non-clast silcrete has the lowest proportion of intensively worked tools in each assemblage except KZ2, where the proportion is slightly higher than that for quartz. The intensity of tool use can be further tested through an examination of the intensity of tool resharpening. One method for doing this is to calculate the mean complete flake to complete tool surface area ratio. This ratio measures the change in flake surface area produced by resharpening in relation to the size of unmodified flakes.

The results for this ratio are presented in Table 8.39 and summary statistics are presented in Table 8.40. The values for each raw material are less than one in all assemblages, a result that indicates two things. First, scrapers from each of the raw materials were manufactured on the largest flakes in the assemblages. Second, scraper resharpening is not very intensive. If scraper resharpening were intensive, the complete flake surface area to complete scraper surface area ratio would be closer to zero because a greater proportion of flake surface area would have been removed through retouch. The value for quartz is identical between the four assemblages, while the two types of silcrete show a greater degree of variability. KZ1 is the only assemblage where the values for both silcretes are greater than that for quartz. At KZ2 the value for non-clast silcrete is greater than quartz, however the value for clast silcrete is not. The CN1 and CN3 assemblages have similar values for the three raw material types and clast silcrete and quartz have the same value in both assemblages. The value for non-clast silcrete is slightly higher at CN3 than CN1. In general, silcrete scrapers appear to be most intensively resharpened in the KZ1 assemblage. Non-clast silcrete scrapers at KZ2 are intensively resharpened compared to scrapers from both CN1 and CN3.

Table 8.39 Complete flake to complete tool scraper surface area ratio. Quartz has the same value in each of the four assemblages and in some assemblages the value for quartz is greater than that of non-clast silcrete indicating more intensive resharpening. In general, clast scrapers are the most intensively resharpened and both types of silcrete are relatively intensively resharpened in the KZ1 assemblage.

Material	CN1	CN3	KZ1	KZ2
Clast	0.6	0.6	0.7	0.5
Non-clast	0.4	0.5	0.7	0.8
Quartz	0.6	0.6	0.6	0.6

Mean Number of Retouched Quadrants

Calculating the mean number of retouched edges on tools is another way of investigating the intensity of tool resharpening. Following Dibble's (1984, 1987) reduction model, more intensive tool resharpening will lead to an increase in the number of retouched edges. The number of retouched edges and the type of modification on each edge were recorded using the quadrant system described by Holdaway and Stern (2004).

Table 8.41 indicates that non-clast silcrete scrapers have the highest mean number of retouched quadrants in the CN1, CN3 and KZ2 assemblages, while the value for non-clast silcrete and clast silcrete is equal for the KZ1 assemblage. The values for clast silcrete are greater than quartz in all the assemblages except for CN3, where quartz scrapers have a mean of 1.7 retouched quadrants compared to 1.6 for

Table 8.40 Summary statistics for complete flake and complete scraper surface area.

Assemblage		Clast complete flake (mm^2)	Clast complete scraper (mm^2)	Non-clast complete flake (mm^2)	Non-clast complete scraper (mm^2)	Quartz complete flake (mm^2)	Quartz complete scraper (mm^2)
CN1	Mean	681.9	1216.5	386.8	931.4	553.7	861.5
	S.D.	625.1	670.6	213.3	590.6	352.3	411.7
	N	466	46	153	19	2719	149
CN3	Mean	652.3	1124.1	399.9	860.1	558.2	869
	S.D.	612.5	421.2	215	435.1	379.6	348.3
	N	280	19	73	11	1689	64
KZ1	Mean	730	1037.5	540.3	723.6	534.2	830.4
	S.D.	530.5	406.2	339.1	315.6	338	391.2
	N	398	25	99	18	712	54
KZ2	Mean	548.9	1104.4	409.5	526.6	489.8	795.2
	S.D.	469.5	719	235.4	289.6	302.8	362.1
	N	1492	77	214	36	2424	83

Table 8.41 Mean number of retouched quadrants on complete scrapers per raw material type in the four assemblages.

Material	CN1	CN3	KZ1	KZ2
Clast silcrete	1.9	1.6	1.8	1.9
Non-clast silcrete	2.3	1.8	1.8	2
Quartz	1.5	1.7	1.4	1.7

clast silcrete. Both the CN1 and the KZ2 assemblages have relatively high means for clast silcrete and non-clast silcrete. Additionally, KZ2 has the highest value for quartz; however, the value for CN1 quartz scrapers is lower than the value for CN3. In general, CN3 and KZ1 have the smallest values. For KZ1, this result is not consistent with the complete flake to complete scraper surface area ratio, which suggested that KZ1 scrapers were intensively resharpened. A possible explanation for this anomaly is that scraper resharpening was concentrated on one rather than multiple margins as is suggested by the results in Table 8.41.

Summary

Larger proportions of non-clast flakes are retouched into tools than flakes of clast silcrete and quartz. This result is consistent with other studies from western New South Wales (Holdaway *et al.* 2004, Shiner 1999). Proportionally, clast silcrete flakes are generally converted into retouched tools more often than quartz flakes, but this is not always the case as demonstrated by the KZ1 assemblage. Although non-clast silcrete flakes are more likely to be converted into tools than flakes of other materials, these will not always be intensively worked. With the exception of tula slugs, non-clast silcrete tools are generally lightly worked and tool forms are characteristically more formal than those of clast silcrete and quartz. The intensity of scraper resharpening is generally low across each raw material and assemblage.

The low proportion of quartz tools is consistent with the non-intensive utilisation of a local raw material source. The more intensive use of clast and non-clast silcrete is consistent with the more intensive reduction of these two materials as indicated by measures of core reduction. In general, the production of tools demonstrates a number of consistent patterns. As was the case for flake attributes, these consistencies contrast with the variability demonstrated by measures of the intensity of core reduction.

Summary of the Composition of the Pine Point-Langwell Assemblages

A number of the characteristics of the assemblages are apparent following analysis. Both types of silcrete were transported most often to the locations as partially decortified nodules. Clast silcrete was predominantly obtained from outcrops, except at CN3 where a significant amount of clast silcrete was obtained from gibber pavements. Non-clast silcrete at CN1 and CN3 was mostly obtained from gibber pavements, but was more likely to have been derived from outcrop sources at KZ1 and KZ2. Quartz was obtained as rounded nodules from nearby gibber pavements and gravel bars in creeks and transported as cortical nodules. Clast silcrete nodules are larger than non-clast silcrete and quartz nodules. These are all long-term patterns that reflect the overall character of the lithic landscape.

Specialised core forms are rare and core rotation is uncommon. Unifacial cores make up the largest proportion of forms at each location, except KZ1, where the proportion of rotated forms was greater than that of unifacial forms. This suggested a concern with extending the reduction life of cores, possibly indicating more intensive core reduction, a hypothesis that does not fit a straight distance decay relationship. Analysis of core reduction intensity indicates that core reduction was least intensive at KZ1, and most intensive at CN1 and KZ2. However, different measures of core reduction provide variable results, with analysis of non-cortical to cortical core ratios not following the results of the MNF to core ratio for CN1 and KZ2. Clearly there is considerable variability in the movement and reduction of clast silcrete.

Non-clast silcrete cores are generally the most intensively reduced, followed by clast silcrete and quartz. Non-clast cores are on average smaller than quartz and clast silcrete cores. The larger size of clast silcrete cores reflects the large size of nodules transported to the locations. Again measures of reduction provide variable results with the non-cortical to cortical complete flake and non-cortical to cortical core ratios highest for clast silcrete, rather than non-clast silcrete as might be expected. This anomaly may reflect the transport of largely decortified non-clast silcrete nodules, which is to be expected because there are no known local sources of this material.

Core platform preparation is rare and largely confined to non-clast silcrete and to a lesser degree clast silcrete. Clast and non-clast silcrete complete flakes at KZ2 are thicker relative to their surface area than flakes in the other assemblages. This suggests extended core reduction and is consistent with the MNF to core ratio. KZ2 complete flakes are also smaller than those in the other assemblages, and have the largest proportion of bi platforms, again suggesting intensive core reduction. As a measure of the intensity with which raw materials were worked, flake characteristics support the MNF to core ratio.

Proportionally, more non-clast silcrete flakes are retouched into tools, followed by clast silcrete and then quartz, except at KZ1, where there are proportionally more quartz complete tools than clast silcrete tools. The two CN assemblages have proportionally more tools for both types of silcrete than the two KZ assemblages. The KZ1 assemblage has the smallest proportion of complete tools for all raw materials. Scrapers are the most common tool form, while specialised forms are rare. The majority of tula occur on non-clast silcrete, but the largest combined proportion of heavily worked tool forms occurs on clast silcrete.

From these results some general patterns can be identified. Non-clast silcrete is the most intensively worked material and was preferentially selected for the production of certain tool forms. This is expected because non-clast silcrete is more likely to be imported from greater distances than either clast silcrete or quartz and therefore greater emphasis was placed on conserving this material. A considerable amount of intra and inter assemblage variability reflects the nature of the lithic landscape. Local sources have a considerable influence on the composition of the assemblages. This is not always represented as a clear distance decay relationship, but aspects of this are evident. Raw material cannot be discounted as a significant factor influencing the degree of long-term behavioural variability reflected in assemblage composition. But, at the same time the long-term influence of raw material has not resulted in the formation of a pattern in any of the assemblages as clear as that identified at SQ1.

It can be tentatively suggested that reduction was most intensive at KZ2, however there is a great deal of inconsistency in the measures of reduction intensity. This suggests that clear patterns between the assemblages are hard to identify and certainly the assemblages do not meet the predicted patterns as suggested by previous lithic research outlined at the beginning of the chapter. The implications of this variability for understanding the nature of long-term place use histories in the Pine Point-Langwell assemblages are explored in the following chapter.

Chapter Nine: The Temporal Character of Assemblage Formation

Introduction

The occupational chronologies established in Chapter Six suggest that the Pine Point-Langwell assemblages have accumulated over broadly similar periods of time. Hence, it is possible to examine the relative amount of behaviour as represented in the formation of the assemblages and relate this to the long-term histories of place use. Some places were the focus of repeated occupation (greater duration and/or more frequent). Schlanger (1992) describes these as persistent places because they represent long-term repeated use. All such places have a combination of characteristics that attract or discourage occupation. These characteristics, particularly those associated with the physical character of places, do not necessarily remain constant but change in accordance with environmental variability and geomorphic change (Fanning and Holdaway 2001b). Therefore, it should not be assumed that artefact accumulation remained constant over the duration of assemblage formation.

Over the long periods of time represented in the archaeological record changes in demographic organisation may result in variable occupational chronologies, e.g., the abandonment of some areas in favour of others, resulting in discontinuous occupational chronologies (Holdaway *et al.* 2002a). Populations are dynamic and population organisation is a major factor structuring the archaeological record (Shennan 2000:821). This concept has a long history in Australian arid zone studies and has been a major focus of debates regarding late Holocene intensification (e.g., Lourandos 1980). However, rarely have these concepts been applied to the investigation of detailed occupational chronologies from surface deposits.

Occupational chronologies that indicate multiple occupational episodes combined with variability in assemblage composition suggest that the Pine Point-Langwell assemblages represent variable histories of place use. The implications of this variability for investigating the nature of place use histories are explored in this chapter. In addition, a broader regional consideration of place use histories is pursued through a comparison of the Pine Point-Langwell assemblages with three assemblages from the nearby ranges where both chronology and access to potential sources of raw material are known.

Inter-assemblage Variability in the Pine Point-Langwell Assemblages

The Pine Point-Langwell assemblages demonstrate variable composition in raw material representation and the degree of reduction intensity. Some aspects of reduction follow a predicted course. For example, quartz, locally abundant as small cobbles, is the least intensively worked material in all of the assemblages and quartz flakes are very similar in shape. This may reflect either long-term continuity in quartz reduction and/or the exploitation of sources with similar characteristics that constrain variation. Indeed, raw material properties (availability, form and internal structure) do account for a significant amount of variability between the assemblages. This is most clearly expressed as a distance decay relationship in raw material proportions. At the same time there is considerable variability that cannot be explained by a distance relationship alone and other factors arising from the nature of the occupations at each place must be considered.

Long-term Patterns

The Pine Point-Langwell distance decay relationship evidenced in the proportional representation of different raw materials is an example of a long-term accumulated pattern, in a similar way to the consistent pattern of raw material reduction identified in the SQ1 quarry assemblage. This aspect of the assemblages suggests a degree of continuity in raw material transport. From this it may be expected that the assemblages furthest away from silcrete sources will demonstrate the most intensive reduction of silcrete. This was tested in Chapter Eight through the analysis of technological ratios (minimum number of flakes to core, non-cortical to cortical core, and non-cortical to cortical flake). However, results of these tests indicate that there is little consistency in the intensity of raw material reduction between the assemblages. Therefore, a re-evaluation of the significance of the distance decay relationship in the overall composition of the assemblages is warranted.

A distance decay model for the Pine Point-Langwell assemblages assumes a degree of stability in raw material exploitation strategies. This is most clearly seen at SQ1 where a consistent pattern of raw material exploitation over time is indicated. SQ1 is not as variable in composition as the creek valley assemblages. It is dominated by a single raw material (clast silcrete) and does not contain a wide range of tool forms or artefacts associated with other activities (e.g., grinding gear) (Table 9.1 and 9.2).

The redundancy of place use is a further factor influencing the formation of long-term patterns. Landscape features are important in conditioning the way people occupy and use space. In arid New South Wales and for that matter in much of the wider Australian arid zone water is an obvious example of such a feature. Present water availability has a restricted and somewhat unpredictable distribution in western New South Wales. This was also the case at various times during the past (e.g., Sturt 1849). Coupled with this is a longer-term

Table 9.1 Artefact frequency and percentage (in brackets) per raw material type in the SQ1.

Category	Clast	Non-clast	Quartz	Quartzite
Broken Flake	1007 (24.7)	15 (20.5)	11 (21.2)	4 (23.5)
Broken Tool	6 (0.1)	1 (1.4)		
Complete Flake	1651 (40.5)	23 (31.5)	12 (23.1)	4 (23.5)
Complete Split	404 (9.9)	13 (17.8)	3 (5.8)	1 (5.9)
Complete Split Tool	4 (0.1)			
Complete Tool	32 (0.8)		2 (3.8)	
Core	108 (2.6)	1 (1.4)	16 (30.8)	1 (5.9)
Distal Flake	407 (10)	8 (11)	4 (7.7)	3 (17.6)
Distal Tool	8 (0.2)		1 (1.9)	
Hammer Stone	1 (0)		2 (3.8)	2 (11.8)
Medial Flake	113 (2.8)	2 (2.7)		
Medial Tool	1 (0)			
Proximal Flake	278 (6.8)	6 (8.2)	1 (1.9)	2 (11.8)
Proximal Split	58 (1.4)	3 (4.1)		
Proximal Tool	1 (0)	1 (1.4)		
Total	4079 (100)	73 (100)	52 (100)	17 (100)

variability arising from environmental change (e.g., Cupper 1998). It would be difficult for people to camp for long periods of time on the residual hillslope at SQ1, where there are no significant water systems. It is likely that a smaller range of behaviours were performed at SQ1 in comparison to the creek valley locations, where occasional water supplies could sustain more prolonged, but not necessarily more regular, occupation. Prolonged occupation increases the opportunity for a wider range of behaviours to be undertaken at a place. Therefore, the creek valley assemblages represent a wider range of behaviours and a high degree of variability. Significantly, these occupations are not of sufficient duration or regularity for the variable behaviours to be swamped by material from a long-term behaviour.

Table 9.2 Complete tool frequency and percentage (in brackets) per raw material type in the SQ1 assemblage.

Tool Type	Clast	Quartz
Denticulate	5 (15.6)	
Notch	5 (15.6)	
Scraper	22 (68.8)	2 (100)
Total	32 (100)	2 (100)

Variable Patterns: Understanding Raw Material Management

Despite the apparent distance decay relationships identified in the study area there is much variability. As noted by Bird (1985), raw materials and artefacts are not always transported in straight lines. In the Pine Point-Langwell case, there is no solid evidence that silcrete from local sources is represented in the creek valley assemblages. It is highly likely that much of the silcrete, especially the non-clast form, was brought into the area from sources located elsewhere. Therefore, the creek assemblages represent multiple stages of raw material reduction. The key to understanding the nature of the variable patterns in the creek valley assemblages is to identify the factors underlying the context of artefact discard.

As outlined in the beginning of Chapter Eight there are two prominent ways to approach the interpretation of assemblage composition. Kuhn (1995) develops a long-term stable strategy to discuss assemblage composition in terms of raw material management strategies. For example, he links provisioning strategies to logistical mobility. Under conditions of increased mobility, strategies of provisioning individuals will predominate because it is difficult to predict the locations where the greatest need for stone artefacts will occur. Under decreased mobility, where the location of future occupation can be predicted with more certainty, a strategy of provisioning locations with raw materials may predominate. These strategies assume that people behave the same way through time. Some aspects of raw material management in the Pine Point-Langwell study area, for instance the representation of raw materials in relation to distance from source support this hypothesis, but others do not. Kuhn's hypothesis is based on the notion that accumulated assemblages represent records of central tendencies but he still assumes that the temporal dimension of the behaviour represented in the record is commensurate with anthropological and socio-ecological theories.

An alternative is to develop an understanding of the temporal character of occupation and then investigate the nature of raw material management. As Kuhn (1995:25-26) points out, mobility and provisioning are not mutually exclusive. It is likely that elements of many different types of provisioning strategy are represented in assemblages, especially if mobility levels varied between occupations. Therefore, technological organisation in palimpsest assemblages is not an either/or, but rather a combination of multiple processes through time.

The degree to which behavioural uniformities can be imposed on the record must be investigated, rather than assumed. Analysis of the Pine Point-Langwell assemblages demonstrated significant variability in a number of technological indices. This variability reflects the nature of raw material management. Despite this, raw material proportions in the four Pine Point-Langwell creek valley assemblages suggest a clear distance decay relationship. Following this, the reduction of silcrete would be expected to be most intensive in the CN1 and CN3 assemblages and less so in the KZ1 and KZ2 assemblages because of their greater distance from known silcrete sources. At CN1 and CN3 there should be a larger number of flakes per core and a greater number of non-cortical artefacts compared to cortical artefacts. However, this is not the case. Instead, there is significant variability in technological ratios irrespective of the location of the assemblage relative to potential raw material sources. This suggests that elements of many different raw material management strategies are represented in the assemblages.

If there was a consistent and significant difference in the duration of occupation between one and more of the Pine Point-Langwell sampling areas this may also influence the types of raw material management strategies implemented. In some contexts raw materials may be cached at locations in anticipation of future use, but at other times raw materials may have been reused or procured under different strategies. Extended occupation might provide more opportunities for the direct procurement of non-local materials because visits to these sources can be planned. Equally, short occupation would limit the time available to utilise local sources, with a greater reliance on transported material. Therefore, procurement is a key element of raw material management that is also linked to the occupational history of locations.

Identifying such signatures in assemblages is challenging, but they are represented in a number of technological measures of raw material management. One such example is the proportion of non-cortical to cortical artefacts. This will vary depending upon the type of raw material management strategy. For example, large proportions of cortical flakes in assemblages are indicative of direct procurement strategies, particularly if these are represented in materials not available within the immediate vicinity of the location. Variability in reduction indices, such as those identified in Chapter Eight may suggest that no one strategy of raw material management dominated over the period represented in assemblage accumulation.

Occupation and Assemblage Accumulation

The wealth of literature discussing the relationship between occupation duration and assemblage composition was summarised in Chapter Three. In brief, as the duration of occupation increases, it is expected that the reduction of raw materials becomes more intensive and the reliance on local materials greater. These patterns are not clear in the Pine Point-Langwell assemblages. Some aspects of assemblage composition are suggestive of greater occupation intensity. For instance, the large number of groundstone artefacts recorded at KZ2 (Table 9.3) could be taken to represent site furniture, cached in anticipation of future return and perhaps indicating extended duration stays (Gould 1980, Kimber and Smith 1988). Alternatively, as demonstrated by Bird (1985), the processes in which artefacts become incorporated into the archaeological record are complicated, and it can be difficult to predict the location of artefact discard.

Artefact discard in the Pine Point-Langwell assemblages certainly conforms to this notion of unpredictability. It is difficult to confidently state that one assemblage represents a series of occupations of greater intensity or duration than another. There are differences in the proportion of tula slugs in the assemblages. Following the probability model of discard, assemblages with larger proportions of tula slugs

Table 9.3 Number of seed grinding artefacts recorded per assemblage for the Pine Point-Langwell study area. Milling slab fragments are the most numerous forms. Also of note is the large number of mortars in the KZ2 assemblage.

Type	CN1	CN3	KZ1	KZ2	Total
Millingslab Fragment	10	18	1	25	54
Mortar	1	1	1	9	12
Muller	4	5	1	4	14
Muller Fragment				1	1
Total	15	24	3	39	81

might represent longer occupational histories compared to assemblages with relatively few tula slugs (Shiner *et al.* 2005). However, this argument is difficult to sustain in the Pine Point-Langwell situation because the overall number of tula slugs in the four assemblages is very small (20 from a total of 28,303 artefacts).

The most likely scenario is that each of the Pine Point-Langwell assemblages reflects a number of occupations of variable duration. Overall, the low intensity of tool use suggests that the majority of occupations were mostly of short-duration. However it is difficult to identify a central tendency in occupation duration because of the variable (inconsistent) results for measures of core reduction intensity between the assemblages and the generally low proportion of heavily worked tools (e.g., tula slugs). It appears that occupation was not intensive enough for a single dominant pattern of place use to emerge.

Two factors need to be considered to understand the relationship between assemblage composition and occupation duration. The first is time since this is one of the most important factors influencing the composition of assemblages. At SQ1, although the temporal context of the place is poorly understood, the geological context suggests that sufficient time had past for repeated occupations to produce a dominant pattern of raw material management. However, this is not the case for the creek valley assemblages, where an insufficient amount of time has past for the formation of a dominant pattern of non-local raw material management. The occupational chronology for the Pine-Rantyga Creek confluence indicates episodic periods of hearth construction during the last 2000 years BP. The erosional context of the assemblages means that direct temporal relationships cannot be drawn between specific hearths and stone artefacts, however it is not unreasonable to expect that many of the artefacts were deposited during the episodes of occupation indicated by the hearths. The episodic nature of occupation combined with the relatively short chronology restricted the formation of a dominant pattern in assemblage composition to form.

The second factor involves the wider landscape context within which each place is situated. This can influence the composition of assemblages in two main ways. First, the environmental composition of the landscape will to a certain extent regulate the types of behaviours performed. For instance, in an arid environment people are unlikely to stay for extended periods of time in areas with no/or poor water resources. In variable landscapes, compositional differences are expected in widely spaced assemblages (e.g., Barton 2001, Veth 1993). In uniform landscapes, it is often assumed that there will be similarities in assemblage composition. However, this scenario is not the case at Pine Point-Langwell. Despite the close proximity of the assemblages and the uniform modern landscape there is still significant intra and inter variability between the assemblages.

Second, as the characteristics of the environment change, the nature of behaviour performed with also change at multiple spatial and temporal scales. Environmental oscillation may be one possible explanation of the episodic nature of hearth construction indicated by the Pine Point-Langwell occupational chronology. Available palaeoenvironmental data suggest a variable climate in southwest New South Wales during the late Holocene. Additionally, correlation between occupational hiatuses and environmental oscillation has been suggested for other areas of western New South Wales (Holdaway *et al.* 2002a, 2005).

The Pine Point-Langwell assemblages display a number of variable characteristics that are difficult to reconcile with one dominant pattern of place use. This may result from a series of occupations of variable duration. Within the assemblages there are also some characteristics that reflect long-term continuity. The consistent reduction of quartz between the four assemblages is a prime example, one that can be explained in terms of the constraints imposed by both the physical characteristics of the raw material and its ubiquitous distribution. Quartz artefacts demonstrate a strong pattern of expedient reduction and use throughout the period represented in the accumulation of the assemblages. Some long-term patterns are also clear in technology and tool typology. All of the assemblages reflect uni-polar reduction and have broadly similar representations of tool forms.
However, the intensity of silcrete reduction (both clast and non-clast) is a major area of difference between the assemblages. While the proportion of raw materials indicates a clear distance decay relationship, there is little evidence that this is supported by consistent increases in core reduction intensity. Some of this variability may be linked to the discontinuous nature of occupation in both of the alluvial systems. Additionally, some may also reflect artefact recycling and secondary transport. In summary, the Pine Point-Langwell assemblages represent complex relationships between many different behavioural processes. In some cases this has resulted in consistent patterns and in others not.

Discussion: Resolution and Behaviour

Stern *et al.* (2002) note that some researchers in East Africa hypothesise that fine resolution deposits offer the best opportunity of identifying clear correlations between material and behaviour. They are sceptical of this view, pointing out that even deposits with fine temporal resolutions represent multiple behavioural events. Therefore, accumulated assemblages are not analogues for single ethnographic events. This is likely to be the case for deflated surface assemblages from western New South Wales since hearth dating indicates multiple episodes of occupation.

Considerable intra assemblage variability indicates that the Pine Point-Langwell assemblages are not representative of a single behaviour. This helps explain the variable intensity of non-clast silcrete and clast silcrete core reduction and may reflect movements of people and materials across the landscape. Raw materials were not always carried to the locations from the same sources. Some materials such as non-clast silcrete were transported in from sources outside the study area. Highly utilised artefacts e.g., tula slugs may have been through numerous stages of reduction and use at other locations before being discarded at one of the four assemblages discussed here. The life history of highly utilised artefacts represents a greater investment of time in manufacture and use when compared to non-intensively utilised artefacts, and they are likely to remain in the cultural system longer. There is also the potential for secondary reuse following initial discard, resulting in further transportation of artefacts between multiple locations.

The variable patterns identified in the Pine Point-Langwell assemblages raise some interesting questions about interpretative scales. Depending upon the perspective adopted, the Pine Point-Langwell assemblages may either represent accumulation over too long a time scale or over to short a time scale. Those seeking ethnographic scale interpretations may argue that the Pine Point-Langwell assemblages are coarse grained with few clear behaviours represented. Others, who are interested in clear average patterns of assemblage composition representative of the dominant types of behaviour may contend that the assemblages do not represent enough accumulation through time, hence variable and non-consistent intra assemblage patterns. Alternatively, it is argued that variable patterns are just as interesting as consistent ones. It is important that an understanding of the temporal context of assemblage accumulation is available, because time, as well as a range of contextual factors influences the formation of either consistent or variable patterns.

Summary: Occupation Histories

The variable nature of the Pine Point-Langwell assemblages makes clear patterns in occupation intensity and duration difficult to detect. It is the differences between the assemblages and within the assemblages as measured by the inconsistent results of core reduction analyses that reflect the nature of occupation at each location.

The discontinuity of occupation indicates that the Pine Point-Langwell assemblages are not the residues of a single adaptive system. However, despite the discontinuous character of occupations, some factors resulted in the formation of clear and consistent patterns within and between the assemblages. The reduction and utilisation of quartz is an obvious example, the implications of this will be discussed in the following chapters. Importantly the results of the Pine Point-Langwell study highlight the danger of assuming rather than demonstrating the repetition of consistent behaviour at any single place through time. Hence the discrepancy between the formation of the record and the application of ethnographically derived interpretative frameworks.

Regional Comparisons: Assemblage Composition and Place Use History at Pine Point-Langwell, Fowlers Gap and Burkes Cave

Demonstrating variable occupational histories for the Pine Point-Langwell assemblages is a positive step towards developing diachronic models of human behaviour from deflated surface assemblages. Clearly, the Pine Point-Langwell assemblages cannot be interpreted using ethnographically derived concepts of occupation intensity, duration and mobility. Whether this pattern is true of other areas of western New South Wales remains to be seen. The inconsistent nature of intra assemblage composition from locations studied on the plains south of the Barrier Range may reflect a unique characteristic of the physical environment, such as the lack of prominent features like waterholes and gorges to attract occupation. This may be investigated through a comparison of the Pine Point-Langwell data with three assemblages from similar chronological contexts in western New South Wales rangeland environments. The assemblages considered here are Burkes Cave (BC), Nundooka (ND) and Sandy Creek (SC). The location of these relative to the Pine Point-Langwell study area is presented in Figure 1.1 and the context of each location is described below.

Description of Burkes Cave (BC)

The stratified open deposit of Burkes Cave is situated on an alluvial terrace in an upland valley of the Scopes Range, 90 kilometres east of Broken Hill. The Scopes Range is composed of an upper-Cambrian quartz rich sequence of sandstone and conglomerate (Mills and Buckley 2000). The eastern flank of the range consists of a basal quartzite pebble/boulder conglomerate overlain by quartzose fluvial sandstone (Mills and Buckley 2000). Siliceous Tertiary deposits, including silcrete, border the range in the northeast. Residual and colluvial deposits of angular poorly sorted sand and gravel dominate the plains to the east of the range, with occasional outcrops of Tertiary silicified conglomerate, quartzite and porcellanite. Flat to gently undulating Quaternary plains of red and brown clayey sand, loam and lateritic soils occur to the west of the range. The plains to the north and south consist of Quaternary dune deposits of red and brown clayey sand, loam and lateritic soils with irregular deposits of aeolian sand. Elevations vary between 354 metres above sea level at Mount Scope to 140 metres above sea level on the surrounding plain. Permanent natural sources of water are confined to waterholes, several of which are located approximately 150 metres west of Burkes Cave in a small rocky channel.

The Scopes Range contains a variety of raw materials suitable for the manufacture of stone artefacts. Quartz occurs as nodules within the sandstone conglomerate that makes up much of the Scopes Range, it also occurs as extensive quartz gibber plains throughout. The exterior of nodules from both contexts is covered in a smooth rounded cortex. There is no quantitative data for the size range of these nodules, but visual inspection indicates that few nodules are larger than the volume of a fist. Quartzite nodules are also widely available, within creek beds and in the gibber pavements on valley slopes. The exterior of nodules from both contexts is covered with a smooth rounded cortex. Silcrete outcrops are restricted to the eastern flank of the Scopes Range.

Harry Allen recovered the assemblage during fieldwork towards his doctoral degree in the late 1960s and early 1970s. A 2.88 square metre area was excavated to a maximum depth of 1.7 meters on the terrace immediately outside a small rockshelter (Allen 1972:143-144). The deposit was excavated in arbitrary 15cm spits that were occasionally modified to fit in with the natural stratigraphy where this could be defined. Four soil horizons reflecting subtle changes in texture, colour and carbon content were defined. A single uncorrected radiocarbon determination of 1850 ± 240 years BP (ANU 704) was obtained on charcoal from a depth of 61 centimetres. This is immediately below the layers that contain the majority of stone artefacts (Allen 1972:150).

The assemblage is currently housed at the Australian Museum in Sydney, where the author and Dr. Simon Holdaway reanalysed the assemblage during May 2001. A total of 6409 artefacts greater than 20 millimetres in maximum dimension were recorded from the BC1/1 and BC1/2 excavation squares.

Description of Fowlers Gap (ND and SC assemblages)

The Fowlers Gap Arid Zone Research Station is situated 110 kilometres north of Broken Hill in the northern hills of the Barrier Range. The ranges to the north, south and west of Fowlers Gap consist of massive sandstone ridges with steep bevelled quartzite ridges and footslopes. Rolling undulating upland plains with extensive quartz gibber pavements occur in the north. To the east of the ranges are scalded Quaternary alluvial plains forming part of the Bancannia Trough, an internal drainage system with no external outlet. Residual and colluvial deposits of sand and gravel are also found on the plain. There are numerous ephemeral creeks draining the range that are characterised by narrow sandy floodplains. Many of these terminate as floodouts on the Bancannia Trough.

The Western New South Wales Archaeology Programme (WNSWAP) conducted three field seasons at Fowlers Gap between 1999 and 2001, recording a total of seven stone artefact assemblages. Two of these assemblages, ND from Sandy Creek in the north of the property and SC from Niela Creek in the south are considered here.

Fowlers Gap represents a rich lithic landscape, with four main types of stone suitable for artefact manufacture. These are quartz, quartzite, sandstone and silcrete. Quartz occurs as outcrops associated with the Barrier Range and as gibber pavements. Numerous quartz outcrops were noted in the nearby Coko Range, however no substantial evidence of on-source reduction was found. Extensive gibber pavements are common and throughout Fowlers Gap. No quantitative data is available, but these generally consist of fist-sized nodules of varying quality and range in colour from milky to opaque. A smooth rounded surface covers the exterior of the nodules. Quartzite is widely distributed and is associated with the Fowlers Gap beds of the Gap Hills and the Nundooka sandstone formation. This material is orthoquartzite (sandstone composed of 90-95% quartz grain) and has a brittle structure not suitable for the manufacture of stone artefacts. Quartzite also occurs in the Teamsters Creek geologic beds to the southeast. The potential of this for stone artefact manufacture is unknown. Silcrete outcrops have been noted in two main areas. The eastern flank of the Gap Hills bordering the alluvial plain contains several isolated, but large outcrops of silcrete with evidence of stone artefact manufacture. Clast size varies from coarse grained to matrix dominated, although medium clast material is by far the most common. Five worked outcrops of silcrete have been recorded on ridge tops in the northern part of the Nundooka formation near Sandy Creek. The composition of the clast size is similar to that of the Gap Hill outcrops. Sandstone suitable for the production of ground artefacts is present in the Nundooka formation, which extends along a north-south axis through the middle of Fowlers Gap and is dominated by massive sandstone ridges intersected by narrow drainage channels.

The ND location consists of an eroding terrace situated on the right bank of the Sandy Creek Gorge. The gorge retains stratigraphic evidence of occasional catastrophic floods that scour sediments from the gorge floor. Radiocarbon determinations to between 3390 and 1710 years BP (Jansen 2001:238) bracket the most recent flood of this type. Fine-grained slackwater deposits draping high terraces along the margins of the gorge began accumulating at least 800 years ago (Jansen 2001:239). These correlate with dates of 709±128 years BP and about 930 years BP (Jansen 2001) obtained from radiocarbon determinations on charcoal from heat retainer hearths excavated on two of these high terrace surfaces. The age of the surface stone artefact deposits at ND can be estimated by combining the results of geomorphological studies of surface age together with a further eight determinations obtained from heat retainer hearths. Both analyses indicate the ND assemblage accumulated over a period of no longer than 1000 years.

The ND assemblage consisting of 4137 artefacts was recorded on the high eroding terrace surfaces. The occupational chronology for ND is based upon eight radiocarbon determinations on wood charcoal from heat retainer hearths (Table 9.4). These indicate that hearth construction occurred approximately between 280 years BP and 840 years BP.

Table 9.4 Radiocarbon determinations on charcoal from heat retainer hearths at the ND location. (Data provided by Dr. Simon Holdaway and Dr. Patricia Fanning)

Hearth ID	Lab ID	DC13	%Modern	Result (year BP)
ND 10	Wk-9339	-22.9 ± 0.2	96.2 ± 0.5	314 ± 39
ND 9	Wk-9338	-21.8 ± 0.2	95.7 ± 0.5	354 ± 39
ND 8	Wk-9337	-23.9 ± 0.2	95.6 ± 0.6	357 ± 53
ND 15	Wk-9342	-24.8 ± 0.2	95.4 ± 0.5	379 ± 46
ND 17	Wk-9343	-23.2 ± 0.2	95.2 ± 1.3	398 ± 111
ND 6	Wk-9336	-23.0 ± 0.2	94.3 ± 0.5	468 ± 38
ND 11	Wk-9340	-24.8 ± 0.2	91.9 ± 0.4	678 ± 37
ND 13	Wk-9341	-24.3 ± 0.2	91.6 ± 1.5	709 ± 128

Table 9.5 Radiocarbon determinations on charcoal from heat retainer hearths at the SC location. (Data provided by Dr. Simon Holdaway and Dr. Patricia Fanning)

Hearth ID	Lab ID	DC13	%Modern	Result (year BP)
SC 11	Wk-10271	-23.8 ± 0.2	97.6 ± 0.5	Modern
SC 34	Wk-10277	-24.5 ± 0.2	97.9 ± 0.5	Modern
SC 36	Wk-10278	-24.3 ± 0.2	97.9 ± 0.5	Modern
SC 81	Wk-9360	-23.3 ± 0.2	96.9 ± 0.5	252±39
SC 52	Wk-9358	-23.6 ± 0.2	96.9 ± 0.6	255±49
SC 10	Wk-10270	-24.0 ± 0.2	96.7 ± 0.6	268±49
SC 23	Wk-10273	-22.8 ± 0.2	96.0 ± 0.6	325±48
SC 44	Wk-9356	-23.7 ± 0.2	95.6 ± 0.4	362±38
SC 27	Wk-10275	-23.9 ± 0.2	95.2 ± 0.6	392±47
SC 28	Wk-10276	-23.6 ± 0.2	93.7 ± 0.5	524±45
SC 69	Wk-10279	-24.4 ± 0.2	92.9 ± 0.7	588±59
SC 13	Wk-10272	-23.6 ± 0.2	89.4 ± 0.5	904±47
SC 24	Wk-10274	-25.5 ± 0.2	84.5 ± 0.7	1349±69

SC is situated on a series of scalded surfaces either side of Niela Creek near its floodout west of the Gap Hills. A discontinuous gallery of river red gum (*Eucalyptus camaldulensis*) presently lines Niela Creek. The creek divides into two separate channels, the active of which is located to the north. A palaeo-channel crosses the present scalded surface to the north of the active channel.

The SC assemblage consists of 4778 artefacts recorded in four major sampling areas on scalded surfaces along the channel. The occupational chronology for SC is based upon 13 radiocarbon determinations on wood charcoal from heat retainer hearths (Table 9.5). These indicate that hearth construction occurred between approximately 200 years BP and 1300 years BP.

Assemblage Comparison

Aspects of the four Pine Point-Langwell assemblages are compared to the assemblages from Burkes Cave (BC) and Fowlers Gap (SC and ND) to provide a wider assessment of assemblage variability across western New South Wales.

Raw Material Representation

A similar range of raw materials to those in the Pine Point-Langwell assemblages is present in the Fowlers Gap and Burkes Cave assemblages, however the proportions of these vary. The proportion of raw materials in each of the assemblages is calculated by counting the number of artefacts per raw material type (Table 9.6). Burkes Cave demonstrates a very different pattern of raw material representation to the other assemblages, being dominated by silcrete as opposed to quartz. ND, SC, KZ1 and KZ2 broadly have similar proportions of silcrete, but quartz is the dominant raw material type. The large proportion of quartz and the small proportion of clast and non-clast silcrete stand out in the CN1 and CN3 assemblages.

Core Reduction Intensity

The minimum number of flakes per core for each of the major raw materials is presented in Table 9.7. The pattern identified for the reduction of quartz in the Pine Point-Langwell assemblages is repeated in the Fowlers Gap and Burkes Cave assemblages with quartz the least intensively worked material. As with the majority of the Pine Point-Langwell assemblages, non-clast silcrete is the most intensively worked material at Burkes Cave, but this is not the case for the ND and SC assemblages, where clast silcrete has the highest ratios, particularly at SC. The ratio for non-clast silcrete at Burkes Cave is significantly higher than KZ2, the next closest. This either suggests that in addition to cores, flakes of this material were transported to the BC locality, or that cores were selectively removed. In summary, the MNF to core ratio indicates considerable variability in the intensity of core reduction for both types of silcrete.

The non-cortical to cortical complete flake ratio (Table 9.8) indicates there are more cortical quartz than non-cortical quartz flakes in all but the KZ1 and KZ2 assemblages. However, this result follows that of the MNF to core ratio (Table 9.7) that indicated the reduction of quartz cores was the least intensive in all assemblages. Again there is a variable pattern for both types of silcrete. The ratio for clast silcrete follows that for the MNF to core ratio in the KZ1, SC and BC assemblages. This is not the case for CN1, CN3 and KZ2, where the non-cortical to cortical complete flake ratio is higher for clast than non-clast silcrete. The same is true for ND, where the non-clast silcrete ratio is higher than that of clast silcrete.

The non-cortical to cortical core ratio (Table 9.9) again demonstrates the low intensity reduction of quartz. In each of the assemblages there are more cortical quartz cores than non-cortical cores. Clast silcrete has the highest ratio in five of the assemblages, CN1, CN3, KZ1, KZ2 and SC.

Summary: Raw Material Reduction

The three measures of core reduction intensity point to a generally variable pattern of raw material utilisation. The consistent pattern of quartz reduction identified at Pine Point-Langwell is repeated in the Fowlers Gap and Burkes Cave assemblages. This may reflect similarities in the form and distribution of quartz in the three study areas. Beyond this there are several aspects of clast and non-clast silcrete reduction that are interesting. The MNF to core ratio for non-clast silcrete is lower than that for clast silcrete at SC, ND and KZ1 confirming that non-clast silcrete is not always the most intensively reduced material. The non-cortical to cortical flake and non-cortical to cortical core ratio support this at both ND and KZ1, but not SC where these ratios are highest for non-clast silcrete. The ND assemblage exhibits an intensity of core reduction similar to that at KZ1. Core reduction is generally intensive at BC and this is supported by a more consistent pattern demonstrated in the non-cortical to cortical core and non-cortical to cortical flake ratios.

Tool Production, Representation and Resharpening

The flake to tool ratio (Table 9.10) shows that quartz has the highest ratio for each of the assemblages. This indicates that proportionally fewer flakes were formally modified into tools in this material. SC is the only assemblage where the ratio for clast silcrete is lower than that of non-clast silcrete. These results indicate that proportionally more tools were made of non-clast silcrete flakes in all assemblages except SC.

Some interesting patterns emerge when the relative intensity of tool production per assemblage and raw material is considered. Overall tool production is not very intensive at BC and KZ2. In fact KZ2 has the highest proportion of flakes per tools for both clast silcrete and quartz. CN1 resides in the middle of tool production intensity. Both CN3 and KZ1 show the lowest ratios for Non-clast silcrete and quartz respectively, however values for the other raw materials reside in the middle of tool production intensity. The two Fowlers Gap assemblages generally show the most intensive tool production, however this does not apply for each of the raw materials. ND has the second lowest values for both clast silcrete and non-clast silcrete, but the value for quartz is relatively high. The low values at ND may mean that few flakes are converted into to tools. The low intensity of core reduction at ND indicates that large numbers of flakes were not produced and therefore flakes are not overwhelming tools. SC has the lowest value for clast silcrete and the second lowest for quartz, but the highest for non-clast silcrete.

The average number of retouched quadrants on complete scrapers per raw material type and assemblage (Table 9.11) indicate that quartz scrapers are the most lightly retouched in each of the assemblage except CN3, where the average for quartz is slightly higher than that for clast silcrete. Non-clast scrapers are generally the most intensively resharpened scrapers. BC, CN1 and KZ2 have high averages for both types of silcrete, and the average for quartz at KZ2 is equal highest with CN1. Non-clast scrapers at ND and SC are relatively intensively resharpened compared to clast scrapers, which have the lowest values. Quartz scrapers at ND and SC are also relatively intensively resharpened.

Discussion: Accumulation and the Formation of the Pine Point-Langwell, Fowlers Gap and Burkes Cave Assemblages

The analysis of the Fowlers Gap and Burkes Cave assemblages confirm some of the general trends in raw material utilisation suggested by the Pine Point-Langwell assemblages. Across

Table 9.6 Raw material proportion and percentage (in brackets) calculated by number for the Pine Point-Langwell, Fowlers Gap (ND and SC) and Burkes Cave (BC) assemblages. Burkes Cave is somewhat different to the other six assemblages being dominated by silcrete as opposed to quartz. CN1 and CN3 have the largest proportion of quartz and the smallest proportion of silcrete.

Material	BC	ND	SC	CN1	CN3	KZ1	KZ2
Clast	2715 (42.2)	1045 (25.3)	970 (20.3)	1091 (12.4)	544 (11.1)	936 (27.4)	3526 (31.5)
Non-clast	1592 (24.8)	654 (15.8)	462 (9.7)	347 (3.9)	145 (3)	248 (7.3)	575 (5.1)
Quartz	1931 (30.1)	2245 (54.3)	3189 (66.7)	7228 (82.2)	4107 (83.7)	2135 (62.4)	6895 (61.6)
Other	171 (2.7)	194 (4.7)	157 (3.3)	122 (1.4)	108 (2.2)	100 (2.9)	196 (1.8)
Total	6409 (100)	4138 (100)	4778 (100)	8788 (100)	4904 (100)	3419 (100)	11192 (100)

Table 9.7 Minimum number of flakes per core for each major raw material class in the Pine Point-Langwell, Fowlers Gap and Burkes Cave assemblages. Non-clast silcrete cores are most intensively worked at BC, KZ2, CN1 and CN3, but clast silcrete cores are more intensively worked than non-clast silcrete cores at KZ1, ND and SC. Quartz cores are the least intensively worked.

Material	BC	ND	SC	CN1	CN3	KZ1	KZ2
Clast	16.8	15.6	22.3	20.8	17.5	16.5	18.3
Non-clast	32.6	14	13.5	23.7	21	14.5	25.4
Quartz	7	5.1	5.3	7.1	7.3	7.3	8

Table 9.8 Non-cortical complete flake to cortical complete flake ratio per raw material and assemblage. Quartz has the lowest ratio for each of the assemblages and only the KZ1 and KZ2 assemblages have proportionally more non-cortical flakes than cortical flakes. Clast silcrete has the highest ratio for the Pine Point-Langwell assemblages, but this does not follow the MNF to core ratio for the CN1, CN3 and KZ2 assemblages.

Material	BC	ND	SC	CN1	CN3	KZ1	KZ2
Clast	5.4	2.2	3	12.3	11.7	9	8.6
Non-clast	10.3	3.1	2.1	9.9	11.2	8.1	7.6
Quartz	0.6	0.7	0.5	0.8	0.2	1.4	1

Table 9.9 Non-cortical core to cortical core ratio per raw material and assemblage. Quartz has the lowest ratio in each of the assemblages (except non-clast silcrete at CN3 where the ratio could not be calculated) indicating that there are more cortical than non-cortical cores. Clast silcrete has the highest ratio in the CN1, CN3, KZ1, KZ2 and SC assemblages. The opposite is true for BC and ND, where the ratio for non-clast silcrete is higher than that for clast silcrete.

Material	BC	ND	SC	CN1	CN3	KZ1	KZ2
Clast	2.5	0.7	1.2	2.6	3.2	3.7	2.1
Non-clast	3.6	1.1	0.8	1.5	0	2.7	1.3
Quartz	0.2	0.2	0.1	0.2	0.1	0.6	0.3

Table 9.10 Flake to tool ratio per assemblage and raw material type. Non-clast silcrete shows the lowest ratio and therefore the most intensive tool production in each of the assemblages except SC. Quartz flakes are less frequently retouched, except for KZ1, where proportionally more quartz flakes are retouched than clast silcrete.

Material	BC	ND	SC	CN1	CN3	KZ1	KZ2
Clast	9.6	7.7	2.5	8.3	10.5	11.7	13
Non-clast	6.7	3.4	7.9	3.8	2.8	4.3	3.9
Quartz	15	16	13.4	13.9	18	9.5	19.9

Table 9.11 Average number of retouched quadrants on complete scrapers per raw material type and assemblage. Quartz scrapers are generally the most lightly resharpened except in the CN3 assemblage, where the average number of retouched quadrants is lower for clast silcrete.

Assemblage	Clast Silcrete	Non-clast Silcrete	Quartz
BC	2	2	1.3
ND	1.6	2.1	1.6
SC	1.7	2	1.7
CN1	1.9	2.3	1.5
CN3	1.6	1.8	1.7
KZ1	1.8	1.8	1.4
KZ2	1.9	2	1.7

the seven assemblages, quartz is consistently the least intensively worked raw material as measured by core reduction, tool production and tool resharpening. Clast silcrete and non-clast silcrete show a variable pattern. Non-clast silcrete cores are generally more intensively worked than clast silcrete cores, but this is not the case for the ND, SC and KZ1 assemblages, where clast silcrete has a higher ratio of flakes to cores than non-clast silcrete. Proportionally more non-clast flakes are selected for tool manufacture than clast silcrete flakes in all but the SC assemblage. Additionally, non-clast silcrete scrapers are also more intensively resharpened than clast silcrete or quartz scrapers. The intensity of scraper resharpening is generally low.

As was the case at Pine Point-Langwell, the Fowlers Gap and Burkes Cave assemblages also indicate that raw material availability was an important factor influencing assemblage composition. This is clearly demonstrated by the Burkes Cave assemblage, where silcrete sources are distant and core reduction intensive. The same applies at SC to a lesser degree. At ND, where worked silcrete outcrops are abundant within one kilometre, cores are not intensively reduced. On its own, the identification of this relationship is not surprising, but it indicates that while a significant amount of the variability between the assemblages can be explained by raw material availability, other aspects of assemblage composition cannot be explained in this way.

BC stands out with the most intensive reduction of non-clast silcrete. It is also located, along with CN1 and CN3, furthest from silcrete sources, so distance to source plays a role in assemblage composition. Significantly though, BC exhibits far more intensive reduction of non-clast silcrete than either CN1 or CN3. From this it is reasonable to conclude that occupations at BC were of longer duration than at any other location considered here. However, this may not be the case if the reduction of other materials is considered. The intensity of silcrete core reduction is not necessarily indicative of the relative duration of multiple occupations. One of the clearest clues to this is the consistency in quartz reduction between the seven assemblages. It is generally expected that non-local raw materials will be more intensively reduced when occupation duration increases (Elston 1990). This results in a greater reliance on local materials, with a corresponding increase in the reduction of these forms. The MNF to core ratio indicates that this is not the case in any of the seven assemblages.

The proportion of different tool forms is also illustrative of the nature of past occupations (Tables 9.12 to 9.14). If intensive non-clast silcrete core reduction at Burkes Cave were correlated with greater duration occupations, then a large proportion of high utility tools, particularly tula slugs would be expected. Comparison of the representation of non-clast silcrete tool frequencies and proportions indicates that the Burkes Cave assemblage does not in fact have the highest proportion of tula and that the proportion of scrapers is reasonably low as well. The ND assemblage has the highest proportion of tula slugs followed by SC and then ND. As discussed previously none of the Pine Point-Langwell assemblages have a high proportion of tula slugs. These results suggest that tool discard is somewhat independent of the intensity of core reduction. The ND assemblage is the best example, with low intensity core reduction yet a large proportion of tula slugs. ND also has a low proportion of non-clast silcrete scrapers. Clearly, there is no simple relationship between core reduction and tool discard in the assemblages considered here. This further highlights the complex formation history of such palimpsest assemblages in western New South Wales.

All these results point to the problematic nature of applying ethnographically derived concepts to the interpretation of material aggregates accumulated over long periods of time. The Fowlers Gap and Burkes Cave assemblages confirm the variable characteristics of occupation as suggested by the analysis of the Pine Point-Langwell assemblages. Although each of the locations reflects variable occupation, Burkes Cave and ND demonstrate very different patterns in assemblage composition. This is in contrast to the Pine Point-Langwell assemblages that demonstrate variable and inconsistent intra assemblage patterns, but ultimately not great inter assemblage differences of the magnitude evident in the ND and Burkes Cave assemblages.

One possible explanation of this may be the wider landscape context within which the assemblages are situated. This has already been discussed for the spatial relationship between potential raw material sources and the locations. The Pine Point-Langwell study area is relatively uniform. Although Pine Creek is a larger channel than Rantyga Creek, the four assemblages are all broadly situated within the confluence of the two channels. The landscape is flat and there are presently no stone lined areas of channel with gorges that may hold water. It is possible that the

Table 9.12 Complete tool form frequency and percentage (in brackets) for clast silcrete per assemblage.

Tool Type	BC	CN1	CN3	KZ1	KZ2	ND	SC
Backed blade	9 (6.8)	1 (1.5)	1 (3.3)		12 (8.7)	5 (6.5)	5 (7.2)
Burin					1 (0.7)		
Burren slug				1 (2.6)	3 (2.2)		
Denticulate	9 (6.8)	9 (13.8)	2 (6.7)	6 (15.8)	14 (10.1)	6 (7.8)	4 (5.8)
Notch	4 (3)	5 (7.7)	2 (6.7)	6 (15.8)	6 (4.3)	4 (5.2)	7 (10.1)
Pirri			3 (10)			1 (1.3)	
Scraper	88 (66.7)	46 (70.8)	19 (63.3)	25 (65.8)	78 (56.5)	40 (51.9)	39 (56.5)
Thumbnail scraper	3 (2.3)	1 (1.5)			2 (1.4)	2 (2.6)	1 (1.4)
Tula slug	4 (3)		2 (6.7)		1 (0.7)	1 (1.3)	1 (1.4)
Utilised	15 (11.4)	3 (4.6)	1 (3.3)		21 (15.2)	18 (23.4)	12 (17.4)
Total	132 (100)	65 (100)	30 (100)	38 (100)	138 (100)	77 (100)	69 (100)

Table 9.13 Complete tool form frequency and percentage (in brackets) for non-clast silcrete per assemblage.

Tool Type	BC	CN1	CN3	KZ1	KZ2	ND	SC
Backed blade				1 (3.6)	1 (1.6)	1 (1.1)	3 (3.5)
Burin	1 (0.8)		1 (4.3)		1 (1.6)	1 (1.1)	
Burren slug	2 (1.7)		1 (4.3)	1 (3.6)	1 (1.6)		
Denticulate	2 (1.7)	2 (4.3)		1 (3.6)	1 (1.6)	6 (6.7)	
Notch	7 (5.8)	3 (6.5)	1 (4.3)	1 (3.6)		7 (7.8)	4 (4.7)
Pirri			1 (4.3)			2 (2.2)	
Scraper	59 (49.2)	19 (41.3)	11 (47.8)	18 (64.3)	36 (58.1)	30 (33.3)	33 (38.4)
Thumbnail scraper	7 (5.8)	1 (2.2)	2 (8.7)		4 (6.5)	4 (4.4)	4 (4.7)
Tula slug	15 (12.5)	7 (15.2)	3 (13)	1 (3.6)	6 (9.7)	21 (23.3)	11 (12.8)
Utilised	27 (19.2)	14 (30.4)	3 (13)	5 (17.9)	12 (19.4)	18 (20)	31 (22.1)
Total	120 (100)	46 (100)	23 (100)	28 (100)	62 (100)	90 (100)	86 (100)

Table 9.14 Complete tool form frequency and percentage (in brackets) for quartz per assemblage.

Tool Type	BC	CN1	CN3	KZ1	KZ2	ND	SC
Backed blade	5 (10.6)	10 (4.3)	8 (7.3)	2 (2.2)	17 (11.6)	2 (3.3)	11 (10.3)
Denticulate	2 (4.3)	15 (6.4)	9 (8.3)	15 (16.9)	18 (12.2)	7 (11.5)	6 (5.6)
Notch	3 (6.4)	43 (18.5)	17 (15.6)	16 (18)	12 (8.2)	3 (4.9)	5 (4.7)
Pirri			1 (0.9)		4 (2.7)		1 (0.9)
Scraper	27 (57.4)	149 (63.9)	64 (58.7)	54 (60.7)	83 (56.5)	29 (47.5)	45 (42.1)
Thumbnail scraper	1 (2.1)	1 (0.4)			4 (2.7)		3 (2.8)
Utilised	9 (19.1)	15 (6.4)	10 (9.2)	2 (2.2)	9 (6.1)	20 (32.8)	36 (33.6)
Total	47 (100)	233 (100)	109 (100)	89 (100)	147 (100)	61 (100)	107 (100)

uniformity of the landscape resulted in little preference for one area over another and this has resulted in the accumulation of assemblages that demonstrate variable behaviour. The assemblages may be best thought of as samples of more extensive artefact distributions that represent many different episodes of occupation in a landscape without obvious geographic barriers to limit the extent and nature of occupation and discard.

The rangeland context of the Fowlers Gap and Burkes Cave assemblages is different to the sand plain of the Pine Point-Langwell study area south of the Barrier Range. Burkes Cave and ND are situated either in or very close to gorges with stone lined creek channels. There are differences in the size of the channels. Sandy Creek is a wide gum tree lined channel, whereas Burkes Cave is situated on a much smaller drainage line. While it is likely that the hydrology of these channels has changed considerably over the last 150 years, these differences are significant enough to have been present during the period in which the assemblages accumulated. Therefore, some of the differences between the ND and Burkes Cave assemblages may reflect the variability of the wider landscape context.

Landscape characteristics are a tempting explanation of assemblage structure and are no doubt an important factor, but they are only one of many. Time is another factor that contributes to variability and it is one that can be investigated, and this is intrinsically related to landscape context. Occupational chronologies have been established for the Pine Point-Langwell and Fowlers Gap assemblages. Unfortunately, only one radiocarbon determination is available from Burkes Cave, but none the less, this provides important information on the commencement of deposit accumulation. Compared to Burkes Cave and the Pine Point-Langwell assemblages, the ND assemblage has accumulated over a relatively short period of time, circa half that represented in the other assemblages.

The chronology at ND is important because it may reflect a shorter period relative to the other locations. Both Burkes Cave and the Pine Point-Langwell locations represent accumulation over at least 2,000 years, while SC represents approximately 1,300 years of potential accumulation. The longer chronologies do not necessarily reflect the duration of individual occupations, but it can be suggested that summed together the individual occupations at each of Burkes Cave, SC and Pine Point-Langwell may be greater in length than the total occupation represented at ND.

Beyond the effects of distance decay on reduction, Burkes Cave indicates relatively short-term occupations, which effectively swamps the assemblage variability reflecting less frequent longer-term occupations. A tendency towards short-term occupations is indicated by the low intensity reduction of locally available quartz. The Pine Point-Langwell assemblages represent a more variable pattern than Burkes Cave, but ultimately show a tendency towards short-term occupations as well. As at Burkes Cave, there is evidence for a distance decay relationship, but this is expressed differently in assemblage composition. Unlike Burkes Cave, the Pine Point-Langwell distance decay relationship is clearly expressed in the proportion of raw materials. Burkes Cave was largely composed of imported materials, while the Pine Point-Langwell assemblages are dominated by local materials with the proportion of non-local silcretes increasing with greater distance from potential sources. Again, unlike Burkes Cave this is not clearly reflected in the intensity of core reduction. Non-local silcretes, especially non-clast silcretes are not as intensively worked as at Burkes Cave, but there are similarities in the intensity of the reduction of locally available quartz. Small proportions of long uselife tools suggest some longer-term occupations, but generally there is a tendency towards periods of much shorter occupation.

It is not clear from the composition of the assemblages whether one place witnessed longer term occupations than others at Pine Point-Langwell. What is indicated is a complex pattern of raw material utilisation, discard and re-use that has resulted in the formation of inconsistent reduction indices. These suggest a different long-term pattern of raw material utilisation between Pine Point-Langwell and Burkes Cave. The Pine Point-Langwell assemblages are characterised by variability, a complex combination of different approaches to the movement, reduction and management of non-local raw materials.

ND demonstrates a different pattern again to the Burkes Cave and Pine Point-Langwell assemblages. The location is characterised by low intensity core reduction suggestive of short-term occupations. Based on this alone, it may be concluded that the place use history of ND is similar to the Pine Point-Langwell assemblages. However, when the representation of high utility tools is considered, it is apparent that the ND assemblage witnessed greater duration occupations more often than the Pine Point–Langwell assemblages despite the shorter span of the occupational chronology. Therefore, the place use history of ND is not like that of Pine Point-Langwell. SC is similar in composition to ND, but may have accumulated over a longer period of time.

Summary

The Burkes Cave assemblage is the least variable, and the vast majority of occupations appear to be of short duration, with occasional episodes of longer duration visits. The Pine Point-Langwell assemblages are variable. The majority of occupations were short duration, but measures of core reduction intensity are rarely consistent between the four assemblages. Although there is a pronounced distance decay relationship represented in raw material proportions, this in not reflected in the intensity of core reduction and the production of clast silcrete tools. It is possible that the sum total of occupations in the Pine Point-Langwell study area was not great enough for a clear pattern in assemblage composition to form. ND is the most historically variable assemblage, with a tendency towards longer duration occupations.

These results have clear implications for the study of settlement patterns, especially for the application of ethnographically derived site types. The seven broadly contemporary assemblages demonstrate varying degrees of historical variability suggestive of different place use histories. Gould (1980:225) suggested that Burkes Cave was one of the richest sites in arid Australia. The re-analysis and comparison of the Burkes Cave assemblage to others with similar chronologies indicates that it is not a semi-permanent base camp. Rather, it is a place that witnessed variable occupation with a tendency towards short duration occupational episodes. Functional site categories can only be used to describe the seven locations if the variability between the assemblages is ignored. The implications of the study are outlined in the next chapter.

Chapter Ten: Conclusion to the Study

Introduction

The potential of deflated surface archaeological deposits as a source of information on past human behaviour in the Australian arid zone is demonstrated in a number of recent papers (e.g., Holdaway *et al.* 1998, Holdaway *et al.* 2000, Holdaway *et al.* 2002a, Fanning and Holdaway 2002, Robins 1996, 1998) and doctoral theses (e.g. Barton 2001, Robins 1993). These contributions have sought to apply various non-site based methodological techniques to the investigation of archaeological deposits that often cover several thousand or more square metres. The present study owes much to these earlier applications of non-site techniques, however, at the same time, it has also expanded upon these studies and developed a different way of addressing the many complex issues pertaining to the study of deflated surface deposits.

As stated earlier, one of the main goals of this research is to highlight the relevance of surface deposits for the investigation of place use histories. The notion of place, as opposed to the site is gaining popularity in Australian archaeology, particularly in cultural heritage management, where there is a growing acknowledgement that the archaeological landscape encompasses complex temporal and spatial configurations of physical remains (Colley 2002). Places with archaeological deposits represent nodes, with contextual characteristics that influenced the nature of artefact discard and the rate of accumulation. Combined with variable geomorphic and environmental conditions, the nature of the behaviours responsible for the discard of artefacts at particular nodes may also change. Therefore, single functional criterions cannot be applied to the interpretation of a particular node, nor can accumulations at two or more different nodes be assumed to represent contemporaneous manifestations of a single adaptive system.

With these in mind, it is possible to re-evaluate the role of recent settlement-subsistence models in the interpretation of the arid zone surface archaeological record. Several prominent models were reviewed in Chapter One (e.g., Thorley 1998b, Veth 1993). This review mainly focused on the construction of occupational chronologies for each of the models. In the present study it is argued that occupational chronologies are crucial to investigating issues of change in settlement-subsistence behaviour and technological organisation. They are also important for understanding the relationship between geomorphic landscape change and the temporal distribution of archaeological materials.

Occupational chronologies indicate that the archaeological record has accumulated over different temporal spans across the landscape. At various temporal scales both continuity and discontinuity can be identified in the accumulation of archaeological materials. At one temporal scale the repeated discard of artefacts and construction of heat retainer hearths over the long-term can be viewed as continuity in the record. However, at the finer temporal scales available within occupational chronologies based on absolute dating, discontinuity can be identified, periods of time where there is little or no evidence for hearth construction and by inference reduced rates of artefact discard. These processes refer to abandonment. Abandonment operates at a range of variable temporal scales (C. Cameron 1993, Nelson 2000, Nelson and Schachner 2002). At the relatively fine temporal resolution afforded by the occupational chronology developed for the Pine Point-Langwell sampling areas a range of discontinuities in hearth construction is identified. These can be viewed as small-scale abandonment events, perhaps ranging in spatial scale from that of an individual location to the wider study area.

The notion of abandonment is an important concept for the establishment of place use histories and has direct implications for interpretations of technological organisation in the context of long-term assemblage accumulation. It is difficult to assess the reasons behind the abandonment of particular parts of the landscape. Nelson (2000) suggests that abandonment may be viewed as processes resulting from social choices within a particular set of circumstances, and that these will create a new set of choices. It is also suggested that the palaeo-environmental record may provide clues as to the possible reasons for long-term abandonment. Environmental factors may also contribute to short-term abandonment. Indeed, short-term abandonment of places and wider landscapes can be viewed as an element of multiple logistical systems. These ideas are explored in relation to the results of the Pine Point-Langwell study.

The Pine Point-Langwell Evidence

Two important data sets were used to investigate the surface archaeology of the study area. The first involved the dating of heat retainer hearths with an emphasis on understanding the relationship between individual hearths and their geomorphic context. The second data set consisted of the analysis of large lithic assemblages from four selected locations where occupational chronologies could be established. Combined, these two data sets permit the investigation of place use history.

Temporal Patterns

The analysis of the radiocarbon determinations in Chapter Six indicated a number of temporal patterns that have implications for the interpretation of the surface archaeological record. All of the 16 radiocarbon determinations from the

Pine Point-Langwell study area fall within the last 2,000 years. This is broadly consistent with radiocarbon determinations from heat retainer hearths in other areas of western New South Wales (e.g., Holdaway *et al.* 2002a, 2005) where the majority of hearths are less than 2000 years old.

The Pine Point-Langwell occupational chronology also indicated phases of hearth construction separated hiatuses with no evidence of hearth construction. Again this pattern is consistent with results from other studies across western New South Wales (Holdaway *et al.* 2002a). It is argued that these hiatuses demonstrate a discontinuity in landscape occupation. Although it is not possible to assign specific artefacts to particular phases of hearth construction, it is argued that the repeated hiatuses have implications for the formation histories of the assemblages. These implications were explored in terms of pattern formation and variation in assemblage composition.

Assemblage Composition

As reviewed in Chapters Eight and Nine, the Pine Point-Langwell assemblages reveal both similarities and differences in structure. These are manifest in raw material reduction intensity, raw material utilisation, artefact discard and the relationship between individual locations within the broader lithic terrain of the study area. What these indicate is that each of the assemblages reflects variable occupational histories at one level and similar processes of accumulation at others. Individually, each of the assemblages is distinct. For instance, the CN1 and CN3 assemblages have very similar physical landscape contexts and although the proportion of different raw material types is very close between them, there are major differences in the degree to which materials were reduced in each assemblage. Based on this fact alone, it was argued that the CN1 location saw a greater intensity of occupation than the CN3 assemblage. Likewise, the KZ1 and KZ2 assemblages demonstrate both similarities and differences in overall assemblage structure, with raw material utilisation more intensive at KZ2.

Consideration of assemblage characteristics indicated that while some of these factors reflected the nature and distribution of raw material sources relative to each of the locations, raw material availability alone could not account for the variable character of the assemblages. Instead, it is argued that the assemblages represent unique occupational histories, with none of them always being occupied in the same way throughout the duration of assemblage accumulation. Indeed, as the occupational chronology from the wider Pine Creek – Rantyga Creek confluence indicates, there are long periods when there is scant evidence for occupation of the area. Over the maximum period of assemblage accumulation, circa 2000 years, it is difficult to sustain the notion of a static behavioural adaptation in the face of variable environmental conditions in southwest New South Wales (see Cupper 1998, Cupper *et al.* 2000). Cupper (1998) notes significant vegetation change during the late Holocene. This is attributed to an increased fire regime resulting from a period of decreased rainfall. However, shorter term environmental variation such as localised drought may also play a significant role. Although not a direct analogue for conditions during the past, historical rainfall records, even over short periods of time indicate considerable variation in the timing and amount of rainfall recorded.

Regional Patterns – Pine Point-Langwell in a Wider Western New South Wales Context

Comparison of the Pine Point-Langwell assemblages with those from other areas of western New South Wales indicates a number of similarities and differences in assemblages that have accumulated over similar temporal units. Many of the similarities reflect common adaptive responses to similar conditions of resource access, particularly the long-term structuring effects of raw material availability. However, similar adaptive responses do not necessarily mean that the locations from which the assemblages were derived formed part of the same settlement/landuse system at any particular period during the past. Indeed, it is likely that many different groups occupying the localities for varied occupation durations deposited archaeological materials.

Differences in assemblage composition between the locations reflect varied occupation histories. Some of these are representative of the variable nature of local lithic landscapes, whilst others are more indicative of the relative duration of individual occupations. Taken as an average, these create a long-term pattern and a trend that can be contrasted between places at multiple temporal and spatial scales.

Landuse Models

Comparison of the Pine Point-Langwell evidence with research undertaken by the Western New South Wales Archaeological Programme in other study areas has enabled a re-evaluation of the notion of settlement pattern systems as they are currently articulated in arid and semi-arid contexts. The establishment and critique of a hypothetical settlement/landuse model for the Pine Point-Langwell study area further illustrates this.

A Model of Landuse for the Pine Point-Langwell Study Area

At this point it is common in Australian arid zone archaeological studies to synthesise the available data into a settlement model. Often these models are extrapolated to regional or pan-continental scales. This approach in part is an artefact of previous studies, dominated by issues regarding the time and nature of colonisation of various ecological zones (e.g., Birdsell 1957, Bowdler 1977, Horton 1981, M. A. Smith 1988, 1989b, Veth 1989). However, more recent approaches derive much from ecological theory, and especially those concerned with optimisation strategies (e.g., Barton 2001).

From the information collected during the Pine Point-Langwell study a settlement system can be proposed. Based upon the observation that places with the most intensive occupation occur adjacent to water sources with the greatest permanency (e.g. Veth 1993:83), different classes of 'sites' are perceived to be located in different parts of the landscape. These are:

- Core Habitation Sites (Long Duration) – adjacent to permanent sources of water and experience the most intensive occupation
- Medium Term Habitation Sites (Medium Duration) – adjacent to less permanent water sources with less intensive occupation than core habitation sites
- Satellite Camps (Short Duration) – adjacent to ephemeral water sources and experience the least intensive occupation

A landuse/settlement model for the Pine Point-Langwell study area based on the above categories can be proposed. Long duration occupation centres (base camps) are predicted to be located adjacent to Pine Creek, the largest creek in the study area and the most likely locality of permanent water sources. Medium duration occupation centres are predicted to be located on less permanent water sources such as the ephemeral and smaller Rantyga Creek. Short duration occupation centres (the equivalent of satellite camps) are predicted to be located in areas that have the most ephemeral water sources, in this case internal drainage basins, box swamps and drainage lines away from the major channels. In addition, the study area contains a range of lithic resources suitable for the manufacture of stone artefacts and the location of these are regarded as an integral component of the settlement model.

Why the Landuse/Settlement Model is Inappropriate

Within this landuse/settlement model it is possible to predict the likely pattern of mobility and resource exploitation strategies employed. For the model to work it is assumed that there is a direct relationship between each of the localities, not at the scale of a yearly mobility/subsistence cycle, but rather over the longer term. Stability is assumed in adaptive responses over the long-term (e.g., Veth 1993:80).

At one level, there is a spatial relationship between all the localities that contain archaeological distributions. However, this does not necessarily imply that there exists an historical or temporal relationship between them. In addition, spatial proximity does not imply that each of the localities functioned the same way within a landuse model over the duration of archaeological accumulation. There are several reasons for this, and each of these is discussed below.

Attempts to reconstruct past hydrological systems from present conditions within the context of settlement systems were criticised in earlier chapters. The main basis of these criticisms stemmed from the significant amount of geomorphic change that has occurred across much of semi and arid western New South Wales during the last 150 years of pastoral landuse (see Fanning 1999). This has resulted in the widespread deposition of sediment and silt in the channels of many streams, in-filling waterholes and other features that may have held water after rains or channel flows. In addition, significant changes have also occurred to the width and depth of channels, with width increasing from erosion caused by the re-direction of stream flow towards the banks by flood debris and other obstacles at the base of the many river red gums that have colonised channel beds (Graeme and Dunkerley 1993).

While it can be hypothesised that a major channel such as Pine Creek had numerous waterholes along its course, it is difficult to know the exact location of these relative to the archaeological distributions that are presently visible on the adjacent land surfaces. There may or may not be a correlation between the two. The instillation of in-ground water storage tanks in drainage systems has also altered hydrological patterns. For example, on Langwell Station, Rantyga Creek no longer flows into Pine Creek, but is dammed, forming a water storage area near the former confluence. Major changes to the organisation of hydrological systems have also been noted prior to the onset of pastoral landuse. At Wannara and Faulkanhagan Creeks east of Broken Hill, Pickard (1994) demonstrated that creek channel changes predate the arrival of pastoral landuse. He concluded that the present changes in hydrological systems merely reflect "the most recent expression of continuing morphological development of these semi-arid creeks adjusting to different combinations of flow and sediment" (1994:281).

Other studies demonstrate the dynamic nature of the semi to arid environment of western New South Wales and these also have serious implications for the construction of behavioural models over the long-term. From the analysis of fossil pollen in the lower Darling Anabranch region, Cupper (1998) discovered major changes in vegetation structure arguing that these reflected increasing aridity or greater seasonal variation in the climate during the late Holocene. One side effect of this was the possibility of an increased incidence of drought. Further studies of Australian rainfall patterns over the twentieth century (e.g. Nicholls and Lavery 1992) indicate considerable variability. Whilst it is dangerous to extrapolate from this to the centuries preceding rainfall records, variable rainfall patterns combined with the results of fossil pollen analysis indicate that the environment of western New South Wales is anything but static over the late Holocene. Changing environmental conditions over a range of temporal and spatial scales have major implications for human landuse strategies. Within this context it is difficult to sustain the notion of a static equilibrium in adaptive systems because changing environmental conditions will prompt variable behavioural responses.

The temporal character of the archaeological record also suggests that the nature of landuse changed. At Stud Creek in northwest New South Wales, people chose not to construct hearths in the valley for a three hundred year period. By implication, it is inferred that use of the valley during this period was less intensive than during times when hearths were constructed. The reasons prompting the change in landuse strategy can only be speculated at however the significance of the discontinuity in occupational chronology resides in the dynamic character of adaptive strategies. Varied landuse strategies may not result in the deposition of different artefact forms or even functional changes in the types of activities performed at individual locations, but rather reflect variability in the scale and tempo of occupation at different places. This variability can take the form of the length of occupation duration or the duration of time between subsequent occupations of locations.

The chronological context of the Pine Point-Langwell deposits indicates periods when there is no evidence for the construction of hearths. This is not to conclude that hearths falling within these periods are not to be found in other parts of the study area. However, within the immediate context of the deposits investigated in this study, there are significant periods of time when hearths do not appear to have been constructed and by implication large amounts of artefacts discarded. This argument is less clear for the KZ1 and KZ2 sampling areas where there are only four determinations available. However, for CN1 and CN3, with a total of 12 determinations, the gaps in the occupational chronology are more convincing.

A further consideration is the age of the surface upon which artefacts and hearths are distributed. The Pine Point-Langwell dates combined with the geomorphic information on surface formation indicate that at a coarse scale, the four surfaces represent occupation occurring over similar and almost contemporaneous temporal units. However, if the finer scale discontinuities in hearth construction are considered, then the nature of occupation at the deposits is anything but contemporaneous. Without establishing occupational chronologies it is impossible to understand the temporal relationship between them different locations. Some of the locations may represent accumulation over significantly greater periods of time than others and therefore provide misleading indications of the relative intensity of occupation.

Assemblage Composition and Assemblage History: A Reconsideration of the Rudall River Data from Western Australia

At this point it is worthwhile reconsidering Veth's interpretation of the relationship between the functional site types and assemblage composition in the Rudall River area. Veth argues for a functional relationship between assemblage composition and the level of water permanency. Assemblages with large proportions of non-local material were argued to represent those assemblages that represent the most permanent occupation. However, a re-analysis of the individual assemblages in Chapter Three demonstrated that there is no clear relationship between increased amounts of non-local materials at locations with greater water permanency. Instead, it is argued that the proportion of raw materials in the assemblages reflects decisions regarding procurement and the form in which nodules were transported. This points to the variable histories of some of the assemblages and suggests alternative explanations as to what they actually represent.

Veth combines all of the assemblages from a single water permanency class for analysis, thus ignoring the possibility of variability between assemblages within a single water permanency class. As noted by Holdaway (1995b, 1995c) combining assemblages in this way makes it difficult to assess the accuracy of Veth's interpretations. At face value the results of the analysis of the combined assemblages suggest a clear correlation between the predictions of assemblage composition and water permanency class. This contrasts to the results of the Pine Point-Langwell study, which indicate significant variability in the intensity of core reduction between assemblages situated within close proximity to one another. Some patterns, such as the reduction of quartz represent one dominant activity, but others like the reduction of the silcretes do not.

The variable pattern identified in the Pine Point-Langwell study is argued to reflect the variable histories of each assemblage. These histories reflect the diverse range of processes that have resulted in the accumulation of palimpsest assemblages. Within this there is both continuity and discontinuity, however no single over-riding factor controls the structure of the assemblages. Veth essentially views the Rudall River study assemblages as functional sets. He argues for a clear distinction between the assemblages based upon their physical location, especially as this relates to the availability and permanency of water resources. The inappropriateness of assuming environmental stability over long periods of time in the arid zone was reviewed in Chapter Four. This factor alone suggests that few, if any assemblages that represent accumulation over the last several thousand years will indicate a stable or single historical record. From this point of view it is unlikely that Veth's amalgamated assemblages represent a single function through time. Instead, it is argued that several of the inconsistencies between the Rudall River assemblages demonstrate the variable nature of assemblage formation.

A re-examination of Veth's analysis indicates several inconsistencies between the combined assemblages. For example, Veth predicts that the proportion of modified flakes will be greatest in assemblages from the permanent water class. However, the tabulation of these indicates that the proportion of modified flakes is largest in the semi-permanent water class, instead of the permanent water class (Veth 1993:36). Further, Veth combines all materials for this analysis thus ignoring the possibility of variation in the proportion of modified flakes between different raw materials within each of the water class categories.

Additional inconsistencies are evident between the assemblages and Veth's predictions for assemblage composition between water classes. Veth (1993:86) predicts that the proportion of whole and broken flakes will be greater at permanent water assemblages. However, the analysis of debitage classes between assemblages from the water class categories indicates that this is not always the case. Unlike the analysis of modified flakes, Veth analyses the proportion of debitage by raw material type. Results indicate considerable variability between the raw materials, with chalcedony and chert following the predicted pattern, but not quartzite and silcrete. Clearly, there is variability in the reduction of different raw materials between the assemblages.

The combination of different raw materials into one analytical category masks variability in the way that raw materials were worked. Therefore, it is difficult to interpret the results of Veth's analysis of modified flake representation because it is not known whether the overall pattern follows that of each raw material type. When analyses are undertaken within individual raw material types variability is detected. This cast some doubt on Veth's interpretation of the assemblages and suggests that variable historical processes were at work.

Veth's model is hampered by small sample size. The total number of artefacts analysed is 12,438. This figure is shared between 14 individual sites, resulting in an average number of artefacts per assemblage of 888.4. As stated in earlier chapters one of the main problems of Australian arid zone lithic studies is the analysis of small assemblages. Clearly, a far more detailed understanding of the characteristics of assemblage composition and formation is required before detailed behavioural interpretations are proposed. Larger assemblages may have also allowed Veth to undertake comparisons between individual assemblages and raw materials.

A Reconsideration of Settlement Systems

There are many problems with recent approaches to the construction of landuse/settlement models in the arid regions of Australia. This study presents a critical appraisal of these. However, at the same time it is acknowledged that the many models proposed for the arid zone follow well established approaches to the reconstruction of past adaptive systems. Additionally, I have emphasised that the outcome of these models is a static adaptive system, in which change occurs infrequently and usually as the result of abrupt and cataclysmic environmental change e.g., large-scale environmental oscillation. However, some researchers (e.g., Lourandos 1996, Ross *et al.* 1992:107) have argued that changes in material culture and adaptive strategies, particularly population increase does not always coincide with environmental change and that these may be indicative of change driven by sociocultural factors. Reducing change to abrupt episodic events is not the desired outcome of the majority of arid zone studies; indeed the stated aim of many of these has been the development of dynamic behavioural models. Therefore, it is appropriate to reconsider the construction of settlement system models in the Australian arid zone.

Continuity/Discontinuity

In this study I have argued for an interpretation of the surface archaeological record that emphasises the dynamic and variable nature of past behavioural systems. This is mainly in response to the inability of previous studies to demonstrate the dynamic nature of such systems. To undertake this, both temporal discontinuities and continuities in the record have been emphasised. Temporal continuities are structures that lead to consistency and similarity in behavioural responses over the long-term. However, as McGlade and van Der Leeuw (1997) note, archaeological interpretations of the long-term are biased towards the emphasis of continuity. In the Pine Point-Langwell study area the distribution and form of raw materials across the landscape was argued to be a major long-term structure influencing the character of archaeological assemblage composition. As well as long-term structures there are consistent patterns in material culture. For example, Shiner (*et al.* 2005) discovered a consistent pattern in flake shape between three geographically separate assemblages in western New South Wales. A similar pattern was discovered for the intensity of quartz core reduction between the Pine Point-Langwell, Fowlers Gap and Burkes Cave assemblages. These similarities may reflect constraints imposed by the physical character of raw materials or approaches to core reduction. What is significant is that these transcend time.

Continuities may be visible over long periods of time in the archaeological record. One of the most interesting recent examples comes from the semi-arid Hay Plain region of southwest New South Wales. Here, Littleton (2002) argued that burial practices on the Hay Plain show a variable but recurrent pattern, which demonstrates continuity over the mid-late Holocene. Regularities in burial practices suggest the maintenance of specific mortuary rules, particularly in the regular spacing of separate burials, which implies an awareness of previous burials. How this pattern relates to landuse strategies is unclear at this stage. Others (e.g., Pardoe 1988) have suggested that burials, particularly when these represent cemeteries have important symbolic properties and that these serve to identify and mark the landscape.

Discontinuities are equally important, but are perhaps more difficult to identify in deflated deposits. On the one hand, these can be seen in occupational chronologies, but the identification of such discontinuities is contingent upon the dating of large numbers of heat retainer hearths. However, few occupational chronologies have been established for deflated deposits. In most studies, occupational chronologies are either derived from dating samples from stratified deposits (e.g., Thorley 1998b, Veth 1993) or they are not established at all (e.g., Barton 2001).

Of course the establishment of occupational chronologies for deflated surface deposits does not allow specific assemblages or groups of artefacts to be correlated with different periods of occupation, either through relative or absolute dating techniques. Deflated assemblages cannot be comparatively analysed, as can those from deposits that retain stratigraphic integrity. The implications of this for identifying changes in material culture, the nature of occupation at individual places and wider landuse strategies were discussed in detail in Chapter Two. To derive dynamic behavioural models from such palimpsest assemblages requires a reconsideration of what change represents in palimpsest archaeological assemblages that are not necessarily historically related. In the context of the archaeological record of the wider western New South Wales area change was argued to represent variability in the relative intensity with which different locations across the landscape were occupied within a given temporal unit of accumulation.

The dynamic character of adaptive strategies can be derived from the comparison of assemblages that have accumulated within different temporal units. The variability in discarded material will not necessarily be reflected in the types of artefacts deposited at locations, but rather will be evidenced in the proportion of different artefacts and the intensity of raw material utilisation. Cane (1984) correctly noted that arid zone assemblages demonstrate considerable variability in the proportional representation of artefact forms, but this variability reflects the nature of occupation span rather that the activities performed at individual locations. Veth also noted this. However, coarse spatial scale and the small size of the Rudall River assemblages precluded the identification of the diverse formational histories of the assemblages.

Resolution is coarsest within palimpsest deposits and the accumulated patterns in assemblage composition reflect the influence of the predominant strategy of occupation, rather than the nature of individual occupations. Occupation of an individual location need not necessarily be of the same character on every occasion; indeed the approach adopted in the present study advocates the opposite view. Instead it is argued that under the variable environmental conditions encountered in western New South Wales, it is unlikely that the nature of occupation at any one location could remain essentially the same over the period over which the archaeological record has accumulated.

Long-Term Place Use Histories: A Dynamic Behavioural Interpretation?

If recent approaches to deriving dynamic adaptive models have been largely unsuccessful, what is the alternative? The surface record of Pine Point-Langwell was investigated as a series of places with individual use histories. These were derived through the investigation of assemblage formation within known occupational chronologies. The results of this analysis reveal variability in assemblage composition across a relatively small area of the landscape. While some aspects of this variability were inferred to represent responses to the distribution and form of lithic raw material sources, for example raw material proportions, others are indicative of variability in the intensity of occupation over the long-term. Different locations exhibited varied occupational signatures. For example, on average occupation intensity at KZ1 and CN3 was less intensive than that at KZ2 and CN1.

On one level, variability of this character is reflective of a landuse system and the argument can certainly be extended that the long-term average pattern of occupation at each individual location reflects the average role of locations within a long-term adaptive strategy. But to adopt this interpretation ignores the variability in the record that is not encompassed within the average model of behaviour. It is likely the nature of occupation and artefact discard is far more complex in organisation than that implied by the comparison of long-term averaged patterns. As consistencies in the archaeological record demonstrate, this variability is not the result of chaotic behaviour because human behaviour is infrequently random, especially in contexts where environmental constraints limit behavioural responses. As Trigger (1991:554) notes "There is enough cross-cultural regularity in widely separated regions to suggest that recurrent operations of cause and effect do result in the evolution of similar forms of behavior to a far greater extent than would be likely to happen as the result of chance alone". The data that does not fit the average model is as interesting as the data that does and demonstrates the flexible and variable character of adaptive systems that existed in the past.

To further demonstrate the potential of place use histories for the development of dynamic models, the analysis of assemblage composition was extended to three other assemblages from additional areas of western New South Wales with established chronological contexts. This comparison revealed both similarities and differences between the three study areas. Quartz demonstrated remarkably similar patterns of reduction, but clast silcrete and non-clast silcrete showed greater variability in strategies of raw material utilisation.

Ultimately, the deflated surface record of western New South Wales documents abandonment. None of the locations investigated in this study represent regular and permanent occupation over the temporal units that the record has accumulated. Within this context abandonment was both planned and un-planned, and this has resulted in the incorporation of artefacts into the archaeological record through varied strategies. For whatever reason, places and even landscapes were abandoned on multiple occasions. It is the dynamic nature of these processes through time that has resulted in the variable character of assemblage composition.

Future Applications for the Investigation of Place Use History in Deflated Deposits

An integrated approach to the investigation of place use history provides an alternative method for the investigation of behavioural signatures in the surface archaeological record. Archaeological and geomorphic data provide the basis for this integration. The attempt at developing these approaches here was limited in scope to four locations in one small area. The Pine Point-Langwell data was then compared to three additional assemblages from other areas of western New South Wales. The main aspect of the approach that marks it as significantly different from its predecessors is a concern with developing occupational chronologies for each location that is investigated. These are crucial to the notion of place use history.

The major future application of the approach is to apply it to the analysis of many more large assemblages with established occupational chronologies. As with current approaches to landuse/settlement models, these assemblages should be derived from a range of different environmental and landscape settings. However, unless these assemblages are investigated within temporal contexts, then effectively the study emulates the a-temporal approaches to settlement/landuse strategies where contemporaniety, even at coarse temporal resolutions is assumed rather than demonstrated (Zvelebil *et al.* 1992). The investigation of many different land surfaces increases the possibility of encountering distributions that have accumulated over different amounts of time. From this perspective, change as variability through time and across space can be investigated, something that has largely been lacking from recent investigations of the arid zone surface record.

Conclusions

Many of the issues covered in this study relate to the scale and resolution at which deflated surface distributions are investigated. Intrinsic to the approach adopted here is the criteria that the history of individual locations be established as occupational chronologies. In many situations it will not be possible to establish this level of chronological control for a number of reasons. In the present study, chronological context could not be established for the SQ1 silcrete quarry beyond the assertion that the quarry is situated on a residual surface that is likely to be much older than that of the scalded surfaces along Rantyga and Pine Creeks. Within this, geomorphological considerations are crucial not only to understanding the present state of archaeological distribution, but also to their formation over the long-term and the likely temporal context over which accumulation has occurred.

Finer temporal control would improve the scale at which continuities and/or discontinuities can be identified in the chronological records. These may point to an even more dynamic history of adaptive systems. However, finer levels of temporal control will not necessarily improve the information potential of deflated palimpsest assemblages. As it stands, there exists no method of correlating individual sets of artefacts with specific occupational episodes. The desired scale of temporal resolution is entirely dependent upon the types of questions being asked and neither fine nor coarse resolutions are necessarily better. In the present case, these questions related to long-term accumulated patterns in discard behaviour and raw material utilisation, therefore a coarse temporal resolution was desired.

The scale at which archaeological phenomena are observed is also important. It was argued that stone artefact and heat retainer hearth distributions should be recorded over large spatial areas, in this case over areas exceeding 5000 square metres. This reflects the low density and dispersed character of artefact distributions that lack clear boundaries (see Holdaway *et al.* 1998). Assemblage size is also an important consideration in lithic studies (see Hiscock 2001). The present study concentrated on recording four large assemblages, with each assemblage exceeding 3000 artefacts. These assemblages are considerably larger than those recorded in other arid zone studies (e.g., Barton 2001, Robins 1993, Thorley 1998b, 1999, Veth 1993) and offer the advantage of providing a large sample of different artefact forms to allow the analysis of variability.

Resolution, scale and the acknowledgement that deflated surface deposits represent palimpsests of material discarded during multiple behavioural episodes formed the backbone of the approach. These concepts were used to develop an alternative approach to the investigation of human behaviour in Australian arid landscapes. In particular, this approach emphasised that the archaeological record is the result of dynamic behaviour processes and that these conflict with the inherently static models that are the result of much archaeological research in the arid zone. This is not to unduly criticise such models, many of these have made significant contributions to our understanding of the arid zone record and the stated aims of these studies has often been the derivation of dynamic behavioural models. Despite this, the interpretations of human behavioural processes generated are static and synchronic. This reflects the methodological and theoretical approaches employed, rather than problems with the record. Ethnography has formed an important component of the construction of adaptive models (e.g., Veth 1993); however it has often been employed substantively rather than methodologically (Bailey 1983). The substantive use of ethnography implies a direct link between ethnographically observed patterns of behaviour and the behavioural processes that have resulted in the discard of archaeological materials. There is an obvious conflict between this approach and the temporal character of the formation of the archaeological record (e.g., Murray 1999, Stern 1993, 1994). The magnitude of this conflict increases when the dynamic nature of the geomorphic and environmental records is considered.

The application of an approach that emphasises place use histories has highlighted the variable character of archaeological assemblage formation. While all the assemblages show variable occupational histories, over the long-term an average pattern of behaviour emerges. In some assemblages, this pattern reflects on average longer term or greater intensity occupations than in others. At the ND locality at Fowlers Gap, the nature of occupation duration appears the most variable and this was argued to reflect the accumulation of this assemblage over a smaller temporal unit than the other six assemblages. This points to the need for a considerable amount of time to pass for the accumulation of a long-term average pattern of occupation character.

The key to extracting behavioural information from deflated assemblages relied on the application of an economic model of assemblage formation that emphasised the relationship between time, discard and lithic raw material utilisation. A number of indices were used to derive this information, in particular the notion that variable levels of mobility result in different constraints that affect the degree of lithic raw material utilisation. However, this does not provide a cookbook recipe for understanding the temporal dimensions of assemblage formation, but rather presents a general guide for understanding the temporal signatures within assemblages. All these factors must be considered within the wider physical context of the location from which each assemblage is derived. In the present study, the availability and the form of raw materials combined with the physical characteristics of each place were argued to be the most important. There has been considerable geomorphic change in western New South Wales and reconstructing the characteristics of past landscapes is not a straightforward task.

This is not to single out water and raw material availability as the only factors influencing the nature of behavioural responses in arid environments. A wide range of complicated factors influences the structure of the archaeological record. Extracting the influence of these in the formation of surface lithic assemblages is difficult; particularly those that are classified as social processes. This has been a criticism of some approaches to arid zone archaeology, claiming that environmental determinism is emphasised at the expense of social factors. However, within the context of the place use histories derived form the present study it is possible to single out some of the long-term behavioural signatures that have structured the archaeological record.

The ultimate test of this approach is the ability to develop an interpretation of deflated surface archaeological deposits that is capable of addressing both issues of continuity and discontinuity. Essentially the end result of this study has been the formulation of a land use/settlement model, but the parameters of the model are not based on those that have preceded it. Future investigations of the arid zone surface archaeological record should consider issues of scale and resolution from a formational perspective. Within the formulation of behavioural interpretations of archaeological phenomena it is all too easy to forget that it the archaeologist who defines the assemblages and the scale of analysis. The assemblages that we study do not have a direct and simple relationship to past human behaviour. Indeed given the dynamic nature of formation it may be entirely inappropriate to use these assemblages to consider the concepts of mobility, land use and occupation duration. Certainly, more consideration should be given to understanding the dynamic characteristics of the record (including temporal context for surface distributions). Towards this objective I have sought to build upon the principles established by the Western New South Wales Archaeological Program. And as demonstrated many years ago with investigations at Lake Mungo, it is at this multidisciplinary level (geoarchaeology and beyond) that future investigations should be undertaken.

There is a clear conflict of temporal scales between models that emphasise the short-term and models that argue that the character of the record reflects long-term processes. Understanding the character of place use histories within temporal contexts and from the perspective of stone artefact manufacture and discard provides a promising avenue of research into the complexities of the surface archaeological record of arid Australia.

References

Adams, W.Y. and E.W. Adams., 1991. *Archaeological Typology and Practical Reality: A Dialectical Approach to Artifact Classification and Sorting*. New York: Cambridge University Press.

Allen, H., 1972. Where the Crow Flies Backwards: Man and Land in the Darling Basin. Unpublished PhD thesis, Department of Prehistory, Australian National University.

Allen, H., 1974. The Bagundji of the Darling Basin: cereal gathers in an uncertain environment. *World Archaeology*, 5:309-322.

Allen, H., 1990. Environmental history in southwestern New South Wales during the Late Pleistocene. In C. Gamble and O. Soffer (eds), *The World at 18 000 BP. Vol. 2: Low Latitudes*. London: Unwin Hyman, pp.296-321.

Allen, H., 1996. Ethnography and prehistoric archaeology in Australia. *Journal of Anthropological Archaeology*, 15:137-59.

Allen, H., 1997. The distribution of large blades: evidence for recent changes in Aboriginal ceremonial exchange networks. In P. McConvell and N. Evans (eds), *Archaeology and Linguistics: Aboriginal Australia in Global Perspectives*, Melbourne: Oxford University Press, pp.357-76.

Allen, H., 1998. Reinterpreting the 1969-1972 Willandra Lakes archaeological surveys. Archaeology in Oceania 33(3):207-20.

Andrefsky, W. Jr., 1994. The geological occurrence of lithic material and stone tool production strategies. *Geoarchaeology*, 9(5):375-91.

Anschuetz, K. F., R. H. Wilshusen, and C. L. Scheick., 2001. An archaeology of landscapes: perspectives and directions. *Journal of Archaeological Research*, 9(2):157-211.

Aplin, K., R. Silcox., N. Stern and E. Williams., 1981. An archaeological survey of the western shore of Lake Cawndilla, Kinchega National Park. In J. Hope (ed), *Darling Surveys 1*. Occasional Papers in Prehistory No. 3. Canberra: Department of Prehistory, Research School of Pacific Studies, Australian National University, pp.71-88.

Bailey, G. N., 1983. Concepts of time in Quaternary prehistory. *Annual Review of Anthropology*, 12:165-92.

Balme, J., 1995. 30,000 years of fishery in western New South Wales. *Archaeology in Oceania* 30(1):1-21.

Balme, J. and J. Hope., 1990. Radiocarbon dates from midden sites in the lower Darling River area of western New South Wales. *Archaeology in Oceania*, 25:85-101.

Bamforth. D. B., and M. S. Becker., 2000. Core/biface ratios, mobility, refitting, and artifact use-lives: a Paleoindian example. *Plains Anthropologist*, 45(173):273-90.

Barbetti, M. and H. Allen., 1972. Prehistoric man at Lake Mungo, Australia, by 32,000 years BP. *Nature*, 240:46-48.

Barton, H. J., 2001. Mobilising Lithic Studies: An Application of Evolutionary Ecology to Understand Prehistoric Patterns of Human Behaviour in the Simpson Desert, far western Queensland. Unpublished PhD thesis, Department of Archaeology, The University of Sydney.

Baumler, M. F., 1988. Core reduction, flake production, and the Middle Paleolithic industry of Zobište (Yugoslavia). In H. L. Dibble and A. Montet-White (eds), *Upper Pleistocene Prehistory of Western Eurasia*. Philadelphia: University of Pennsylvania Museum, pp.255-74.

Binford, L, R., 1977. Forty-seven trips: a case study in the character of archaeological formation processes. In R. V. S. Wright (ed), *Stone Tools as Cultural Markers; Change, Evolution, and Complexity*, Canberra: Australian Institute of Aboriginal Studies, pp. 24-36.

Binford, L.R., 1978. Dimensional analysis of behavior and site structure: learning from an Eskimo hunting stand. *American Antiquity*, 43(3):330-61.

Binford, L.R., 1979. Organization and formation processes: looking at curated technologies. *Journal of Anthropological Research*, 35(3):255-273.

Binford, L.R., 1980. Willow smoke and dogs tails: hunter-gather settlement systems and archaeological site formation. *American Antiquity*, 45(1):4-20.

Binford, L.R., 1982. The archaeology of place. *Journal of Anthropological Archaeology*, 1(1):5-31.

Binford, L.R., 1992. Seeing the present and interpreting the past – and keeping things straight. In J. Rossignol and L. Wandsnider (eds), *Space, Time and Archaeological Landscapes*, New York: Plenum Press, pp.43-59.

Binford, L.R., 2001. Constructing Frames of Reference: An Analytical Method for Archaeological Theory Building Using Hunter-Gather and Environmental Data Sets. Berkeley: University of California Press.

Binford, L. R. and J. F. O'Connell., 1984. An Alyawara day: the stone quarry. *Journal of Anthropological Research*. 40:406-32.

Binford, L. R. and J. A. Sabloff., 1982. Paradigms, Systematics, and Archaeology. *Journal of Anthropological Research*, 38(2):137-53.

Bird, C.F.M., 1985. Prehistoric Lithic Resource Utilisation: A case study from the Southwest of Western Australia. Unpublished PhD thesis, Centre for Prehistory, The University of Western Australia.

Bird, C. F. M., and D. Frankel., 1991a. Chronology and explanation in western Victoria and south-east South Australia. *Archaeology in Oceania*, 26(1):1-16.

Bird, C. F. M., and D. Frankel., 1991b. Problems in constructing a prehistoric regional sequence: Holocene southeast Australia. *World Archaeology*, 23(2):179-91.

Birdsell, J. B., 1957. Some population problems concerning Pleistocene man. *Cold Springs Harbor Symposia on Quantitative Biology*, 22:47-69.

Blades, B. S., 1999. Aurignacian lithic economy and early modern human mobility: new perspectives from classic sites in the Vézère valley of France. *Journal of Human Evolution*, 37:91-120.

Blumenschine, R. J., and F. T. Masao., 1991. Living sites at Olduvai Gorge, Tanzania? Preliminary landscape archaeology results in the basal Bed II lake margin zone. *Journal of Human Evolution*, 21:451-62.

Bourke, M., 1997. The age of two human occupation sites in the Eastern Macdonald Ranges, Central Australia. *Australian Archaeology*, 45:55-57.

Bowdler, S., 1977. The coastal colonisation of Australia. In J. Allen, J. Golson and R. Jones (eds), *Sunda and Sahul*. London: Academic Press, pp.204-46.

Bowdler, S., 1984. Archaeological significance as a mutable quality. In S. Sullivan and S. Bowdler (eds), *Site Surveys and Significance Assessment in Australian Archaeology*. Canberra: Department of Prehistory, Research School of Pacific Studies, Australian National University, pp.1-9.

Bowdler, S., and S. O'Connor., 1991. The dating of the Australian Small Tool Tradition, with new evidence from the Kimberley, WA. *Australian Aboriginal Studies*, 1:53-62.

Bowler, J. M., 1986. Quaternary landform evolution. In D.N. Jeans (ed.), *Australia - A Geography. Vol I: The Natural Environment*. Sydney: Sydney University Press, pp.117-147.

Bowler, J. M., 1998. Willandra Lakes revisited: environmental framework for human occupation. *Archaeology in Oceania*, 33(3):120-55.

Bowler, J. M., R. Jones, H. Allen, and A. G. Thorne., 1970. Pleistocene human remains from Australia: a living site and human cremation from Lake Mungo, western New South Wales. *World Archaeology*, 2:39-60.

Bowler, J. M., A. G. Thorne, and H. A. Polach., 1972. Prehistoric man in Australia: age and significance of the Mungo skeleton. *Nature*, 240:48-50.

Bowler, J. M. and A. G. Thorne., 1976. Human remains from Lake Mungo. Discovery and excavation of Lake Mungo III. In R. L. Kirk and A. G. Thorne (eds), *The Origins of the Australians*. Canberra: Australian Institute of Aboriginal Studies, pp.127-38.

Bowler, J. M., E. Stockton., and M. J. Walker., 1978. Quaternary stratigraphy of the Darling River near Tilpa, New South Wales. *Proceedings of the Royal Society of Victoria*, 90:79-88.

Bowler, J. M., and D. M. Price., 1998. Luminescence dates and stratigraphic analyses at Lake Mungo: review and new perspectives. *Archaeology in Oceania*, 33(3):156-168.

Bowler, J. M. and J. Magee., 2000. Redating Australia's oldest human remains: a skeptics view. *Journal of Human Evolution*, 38:719-726.

Bradley, R., 1991. Ritual, time and history. *World Archaeology*, 23(2):209-19.

Brantingham, P. J., J. W. Olsen, J. A. Rech and A. I. Krivoshapkin., 2000. Raw material quality and prepared core technologies in Northeast Asia. *Journal of Archaeological Science*, 27:255-71.

Braudel, F., 1980. *On History*. London: Weidenfeld and Nicholson.

Brown, A.G., 1997. *Alluvial Geoarchaeology: Floodplain Archaeology and Environmental Change*. Cambridge: Cambridge University Press.

Brown, P., 1987. Pleistocene homogeneity and Holocene size reduction: the Australian human skeletal evidence. *Archaeology in Oceania* 22(2):41-67.

Buchan, R. and J. P. White., 1981. An archaeological survey in Kinchega National Park. In J. Hope (ed), *Darling Surveys 1*, Occasional Papers in Prehistory No. 3, Canberra: Department of Prehistory Research School of Pacific Studies Australian National University, pp.65-70.

Burton, G. R., 1994. Metallogenic Studies of the Broken Hill and Euriowie Blocks, New South Wales, No. 3. Mineral Deposits of the South-eastern Broken Hill Block. Department of Mineral Resources, Geological Survey of New South Wales.

Butzer, K. W., 1994. *Archaeology as Human Ecology*. Cambridge: Cambridge University Press.

Byrne, D., 1980. Dynamics of dispersion: the place of silcrete in archaeological assemblages from the Lower Murchison, Western Australia. *Archaeology and Physical Anthropology in Oceania*, 15:110-19.

Cameron, C. M., 1993. Abandonment and archaeological interpretation. In C.M. Cameron and S. A. Tomka (eds), *Abandonment of Settlements and Regions: Ethnoarchaeological and Archaeological Approaches*. Cambridge: Cambridge University Press, pp.3-7.

Cameron, D., P. White, R. Lampert and S. Florek., 1990. Blowing in the wind: site destruction and creation at Hawker Lagoon, South Australia. *Australian Archaeology*, 30:58-69.

Camilli, E. L., 1983. Site Occupational History and Lithic Assemblage Structure: An example from Southeastern Utah. Unpublished PhD thesis, Department of Anthropology, University of New Mexico.

Camilli, E L., and J. I. Ebert., 1992. Artifact reuse and recycling in continuous surface distributions and implications for interpreting land use patterns. In J. Rossignol and L. Wandsnider (eds), *Space, Time and Archaeological Landscapes*, New York: Plenum Press, pp.113-36.

Cane, S., 1984. Desert Camps: A Case Study of Stone Artefacts and Behaviour in the Western Desert. Unpublished PhD thesis, Department of Prehistory, Australian National University.

Christian, C. S., and G. A. Stewart., 1953. *General Report on Survey of Katherine-Darwin Region, 1946: Land Research Series No.1*. Melbourne: Commonwealth Scientific and Industrial Research Organization, Australia.

Clarke, D. L., 1977. Spatial information in archaeology. In D. L. Clarke (ed), *Spatial Archaeology*. London: Academic Press, pp.1-32.

Clarkson, C., 2002. An index of invasiveness for the measurement of unifacial and bifacial retouch: a theoretical, experimental and archaeological verification. *Journal of Archaeological Science*, 29:65-75.

Close, A. E., 1999. Distance and decay: an uneasy relationship. *Antiquity*, 73:24-32.

Cochrane, G. W., G. W., Quick and D. Spencer-Jones. 1991. *Introducing Victorian Geology*. Melbourne: Geological Society of Australia.

Colley, S., 2002. *Uncovering Australia: Archaeology, Indigenous People and the Public.* Crows Nest: Allen and Unwin.

Cosgrove, R., 1995. *The Illusion of Riches: Scale, Resolution and Explanation in Tasmanian Pleistocene Human Behaviour.* Oxford: BAR Publishing, BAR International Series 608.

Cosgrove, R., 1999. Forty-two degrees south: the archaeology of Late Pleistocene Tasmania. *Journal of World Prehistory*, 13(5):357-402.

Cosgrove, R., 2000. Lithic raw material distribution in Tasmania. In A. Anderson and T. Murray (eds), *Australian Archaeologist: Collected Papers in Honour of Jim Allen*, Canberra: Coombs Academic Publishing, pp.187-200.

Cosgrove, R., J. Allen and B. Marshall., 1990. Palaeo-ecology and Pleistocene human occupation in south central Tasmania. *Antiquity*, 64:59-78.

Cosgrove, R., and J. Allen., 1996. Research strategies and theoretical perspectives. In J. Allen (ed), *Report of the Southern Forests Archaeological Project, Volume 1; Site Descriptions, Stratigraphies and Chronologies.* Bundoora: Archaeology Publications School of Archaeology La Trobe University, pp.21-30.

Cosgrove, R., and J. Allen., 2001. Prey choice and hunting strategies in the Late Pleistocene: evidence from southwest Tasmania. In A. Anderson, I. Lilley and S. O'Connor (eds), *Histories of Old Ages: Essays in Honour of Rhys Jones.* Canberra: Pandanus Books, Research School of Pacific and Asian Studies. pp.397-429.

Cramsie, J. N. and J. M. Hawke 1984. Introduction. In J. M. Hawke and J. N. Cramsie (eds), *Contributions to the Geology of the Great Australian Basin.* Geological Survey of New South Wales Bulletin 31, Department of Mineral Resources, pp.1-22.

Cunningham, G. M., W. E. Mulham., P. I. Milthorpe and J. H. Leigh., 1992. *Plants of Western New South Wales.* Sydney: Inkata Press.

Cupper, M., 1998. Casuarinuceae: Chenopodiaceae pollen ratios and soil salinization in the Murray-Darling Basin. *Quaternary Australasia*, 16(2):28-34.

Cupper, M. L., A. N. Drinnan., and I. Thomas., 2000. Holocene palaeoenvironments of salt lakes in the Darling Anabranch region, south-western New South Wales, Australia. *Journal of Biogeography*, 27:1079-94.

Dancey, W.S., 1998. The value of surface archaeological data in exploring the dynamics of community evolution in the Middle Ohio Valley. In A. P. Sullivan III (ed.), *Surface Archaeology.* Albuquerque: University of New Mexico Press, pp.3-19.

David, B., and H. Lourandos., 1997. 37,000 years and more in tropical Australia: investigating long-term archaeological trends in Cape York Peninsula. *Proceedings of the Prehistoric Society*, 63:1-23.

David, B., I. McNiven, J. Flood, R. Frost., 1990. Yiwarlarlay 1: archaeological excavations at the Lightning Brothers site, Delamere Station, Northern Territory. *Archaeology in Oceania*, 25:79-84.

David, B., and M. Wilson., 1999. Re-reading the landscape: place and identity in NE Australia during the Late Holocene. *Cambridge Archaeological Journal* 9(2): 163-88.

David, N., and J. Sterner., 1999. Wonderful society: the Burgess shale creatures, Mandara polities, and the nature of prehistory. In S. K. McIntosh (ed), *Beyond Chiefdoms: Pathways to Political Complexity in Africa.* Cambridge: Cambridge University Press, pp.97-109.

Davidson, Iain., 1983. On the edge of the Simpson: recent additions to understanding of prehistoric artefact distributions in arid Australia. *Australian Archaeology*, 17:27-37.

Dewar, R. E., and K. A. McBride., 1992. Remnant settlement patterns. In J. Rossignol and L. Wandsnider (eds), *Space, Time and Archaeological Landscapes*, New York: Plenum Press, pp.227-55.

Dibble, H. L., 1984. Interpreting typological variation in Middle Paleolithic scrapers: function, style or sequence of reduction? *Journal of Field Archaeology*, 11:431-36.

Dibble, H. L., 1987. The interpretation of Middle Paleolithic scraper morphology. *American Antiquity*, 52(1):109-17.

Dibble, H. L., 1988. Typological aspects of reduction and intensity of utilization of lithic resources in the French Mousterian. In H. L. Dibble and A. Montet-White (eds), *Upper Pleistocene Prehistory of Western Eurasia.* Philadelphia: University of Pennsylvania Museum, pp.181-97.

Dibble, H. L., 1991. Local raw material exploitation and its effects on Lower and Middle Paleolithic assemblage variability. In A. Montet-White and S. Holen (eds), *Raw Material Economies among Prehistoric Hunter-Gathers.* Lawrence: University of Kansas, pp.33-47.

Dibble, H. L., 1995a. Middle Paleolithic scraper reduction: background, clarification and review of the evidence to date. *Journal of Archaeological Method and Theory*, 2(4):299-368.

Dibble, H. L., 1995b. Raw material availability, intensity of utilization, and Middle Paleolithic assemblage variability. In H. Dibble and M. Lenoir (eds), *The Middle Paleolithic Site of Combe-Capelle Bas (France).* Philadelphia: University of Pennsylvania Museum, pp.289-315.

Dibble, H. L., 1997. Platform variability and flake morphology: a comparison of experimental and archaeological data and implications for interpreting prehistoric lithic technological strategies. *Lithic Technology*, 22:150-70.

Dibble, H. L., and S. P. McPherron., 1988. On the computerization of archaeological projects. *Journal of Field Archaeology*, 15:431-40.

Dibble, H. L., and N. Rolland., 1992. On assemblage variability in the Middle Paleolithic of Western Europe: history, perspectives, and a new synthesis. In H. L. Dibble and P. Mellars (eds), *The Middle Paleolithic: Adaptation, Behavior and Variability.* Philadelphia: University of Pennsylvania Museum, pp.1-25.

Dibble, H. L., B. Roth and M. Lenior., 1995. The use of raw materials at Combe-Capelle Bas. In H. Dibble and M. Lenior (eds), *The Middle Paleolithic Site of Combe-Capelle Bas (France).* Philadelphia: University of Pennsylvania Museum, pp.259-87.

Dickson, F. P., 1977. Quartz flaking. In R. V. S. Wright (ed), *Stone Tools as Cultural Markers; Change, Evolution, and Complexity*, Canberra: Australian Institute of Aboriginal Studies, pp.97-103.

Doelman, T., J. Webb and M. Domanski., 2001. Source to discard: patterns of lithic raw material procurement and use in Sturt National Park, northwestern New South Wales. *Archaeology in Oceania*, 36(1):15-33.

Duke, P., 1996. Braudel and North American archaeology: an example from the Northern Plains. In R. W. Preucel and I. Hodder (eds), *Contemporary Archaeology in Theory: A Reader*, Cambridge: Blackwell, pp.240-57.

Dunkerley, D. L., 1992. Channel geometry, bed material, and inferred flow conditions in ephemeral stream systems, Barrier Range, Western N.S.W. Australia. *Hydrological Processes*, 6:417-33.

Dunkerley, D. L., 1999. Banded chenopod shrublands of arid Australia: modelling responses to interannual rainfall variability with cellular automata. *Ecological Modelling*, 121:127-38.

Dunnell, R. 1980. Evolutionary theory and archaeology. *Advances in Archaeological Method and Theory*, 3:35-99.

Dury, G.H. and T. Langford-Smith., 1970. A Pleistocene Aboriginal camp fire from Lake Yantara, northwestern New South Wales. *Search*, 1(2):73.

Ebert, J. I., 1992. *Distributional Archaeology*. Albuquerque: University of New Mexico Press.

Edwards, P. C., 1995. The formation of large sites and their interpretation: evidence from the Pleistocene archaeological record of the East Jordan Valley. *The Artefact*, 18:60-63.

Elston, R., 1990. A cost-benefit model if lithic assemblage variability. In R. Elston and E.E. Bundy (eds), *The Archaeology of James Creek Shelter*. Salt Lake City: University of Utah Anthropological Papers 115, pp.153-64.

Fanning, P., 1994. Long-term contemporary erosion rates in an arid rangelands environment in western New South Wales, Australia. *Journal of Arid Environments*, 28:173-87.

Fanning, P., 1999. Recent landscape history in arid Western New South Wales, Australia: a model for regional change. *Geomorphology*, 29:191-209.

Fanning, P., 2002. Beyond the Divide: A New Geoarchaeology of Aboriginal Stone Artefact Scatters in Western NSW, Australia. Unpublished PhD thesis, Graduate School of the Environment, Division of Environmental and Life Sciences, Macquarie University.

Fanning, P., and S. J. Holdaway., 2001a. Temporal limits to the archaeological record in arid Western NSW, Australia: lessons from OSL and radiocarbon dating of hearths and sediments. In M. Jones and P. Sheppard (eds), *Australasian Connections and New Directions: Proceedings of the 7th Australasian Archaeometry Conference*. Research in Anthropology and Linguistics 5, pp.85-104.

Fanning, P., and S. J. Holdaway., 2001b. Stone Artifact Scatters In Western NSW, Australia: Geomorphic Controls On Artifact Size. *Geoarchaeology* 16(6):667-86.

Fletcher, R. 1992., Time perspectivism, *Annales*, and the potential of archaeology. In B. A. Knapp (ed), *Archaeology, Annales and Ethno-history*. Cambridge: Cambridge University Press, pp.34-49.

Fletcher, R., 1995. *The Limits of Settlement Growth, A Theoretical Outline*. Cambridge: Cambridge University Press.

Flood, J., 1984. More or less significant: a national perspective on assessing the significance of archaeological sites. In S. Sullivan and S. Bowdler (eds), *Site Surveys and Significance Assessment in Australian Archaeology*. Canberra: Department of Prehistory, Research School of Pacific Studies, Australian National University, pp.55-60.

Florek, S. M., 1987. The archaeological variability of mound springs sites at Lake Eyre South, South Australia: work in progress. *Australian Archaeology*, 24:27-31.

Foley, R., 1981. Off-site archaeology: an alternative approach for the short-sited. In I. Hodder, G. Isaac, and N. Hammond (eds), *Pattern of the Past: Studies in Honour of David Clarke*. Cambridge: Cambridge University Press, pp.157-83.

Frankel, D. 1988. Characterising change in prehistoric sequences: a view from Australia. *Archaeology in Oceania*, 23:41-48.

Frankel, D., 1993. Pleistocene chronological structures and explanations: a challenge. In M. A. Smith, M. Spriggs and B. Fankhauser (eds), *Sahul in Review: Pleistocene Archaeology in Australia, New Guinea and Island Melanesia*, Canberra: Department of Prehistory, Research School of Pacific Studies The Australian National University, pp.24-33.

Fullagar, R., 1986. Use-wear on quartz. In G. Ward (ed), *Archaeology at Anzaas*. Canberra: Canberra Archaeological Society, pp.191-93.

Fullagar, R., 1994. Traces of times past: stone artefacts into prehistory. *Australian Archaeology*, 39:63-73.

Gamble, C., 1993. Exchange, foraging and local hominid networks. In C. Scarre and F. Healy (eds), *Trade and Exchange in Prehistoric Europe*, Oxford: Oxbow Books, pp.35-44.

Gamble, C., 1999. *The Palaeolithic Societies of Europe*. Cambridge: Cambridge University Press.

Gill, E. D., 1973. Geology and geomorphology of the Murray River region between Mildura and Renmark, Australia. *Memoirs of the National Museum of Victoria*, 34:1-97.

Gnaden, D. and S. Holdaway., 2000. Understanding observer variation when recording stone artifacts. *American Antiquity*, 65(4):739-47.

Gosden, C., 1994. *Social Being and Time*. Oxford: Blackwell.

Gosden, C., and L. Head., 1994. Landscape – a usefully ambiguous concept. *Archaeology in Oceania*, 29:113-16.

Gosden, C., and G. Lock., 1998. Prehistoric histories. *World Archaeology*, 30(1):2-12.

Gould, R. A., 1969. Puntutjarpa Rockshelter: A reply to Messrs. Glover and Lampert. *Archaeology and Physical Anthropology in Oceania*, 4:229-37.

Gould, R. A., 1978. Beyond analogy in ethnoarchaeology. R. A. Gould (ed), *Explorations in Ethnoarchaeology*. Albuquerque: University of New Mexico Press, pp.249-93.

Gould, R. A., 1980. *Living Archaeology*. Cambridge: Cambridge University Press.

Gould, R. A., 1991. Arid-land foraging as seen from Australia: adaptive models and behavioral realities. *Oceania*, 62:12-33.

Gould, R. A., 1995. A review of: Islands in the Interior: The Dynamics of Prehistoric Adaptations within the Arid Zone of Australia. *American Antiquity*, 60(3):562-63

Gould, R. A. and S. Saggers., 1985. Lithic procurement in Central Australia: a closer look at Binford's idea of embeddedness in archaeology. *American Antiquity*, 50(1):117-36.

Graeme, D., and D. L. Dunkerley., 1993. Hydraulic resistance by the river red gum, *Eucalyptus camaldulensis*, in ephemeral desert streams. *Australian Geographical Studies*, 31(2):141-54.

Griffiths, T., 1996. *Hunters and Collectors*. Cambridge: Cambridge University Press.

Hale, H. and N. Tindale., 1930. Notes on some human remains in the lower Murray Valley, South Australia. *Records of the South Australian Museum*, 4(2):145-218.

Hall, J., and W. R. F. Love., 1985. Prickley Bush, a site with backed blades on the Brisbane River: a pilot study towards the measurement of site "disturbance", *Queensland Archaeological Research*, 2:71-81.

Hallam, S. J., 1985. Microlithic industries in Western Australia: some aspects. In V. N. Misra and P. Bellwood (eds), *Recent Advances in Indo-Pacific Prehistory: Proceedings of the International Symposium Held at Poona, December 19-21, 1978*. New Delhi: Oxford and IBH Publishing Co, pp.219-29.

Hawkins, R. A., 1998. Coming full circle: plowzone assemblages and the interpretation of Fort Ancient settlement structure. In A. P. Sullivan III (ed.), *Surface Archaeology*. Albuquerque: University of New Mexico Press, pp.91-109.

Hayden, B., 1977. Stone tool functions in the Western Desert. In R. V. S. Wright (ed), *Stone Tools as Cultural Markers; Change, Evolution, and Complexity*, Canberra: Australian Institute of Aboriginal Studies, pp.178-88.

Hayden, B., 1979. *Palaeolithic Reflections: Lithic Technology and Ethnographic Excavations Among Australian Aborigines*. Canberra: Australian Institute of Aboriginal Studies.

Hayden, B., N. Franco and J. Spafford., 1996. Evaluating lithic strategies and design criteria. In G. Odell (ed), *Stone Tools: Theoretical Insights into Human Prehistory*. New York: Plenum Press, pp.9-45.

Hemming, S., V. Wood and R. Hunter., 2000. Researching the past: oral history and archaeology at Swan Reach. In R. Torrence and A. Clarke (eds), *The Archaeology of Difference: negotiating Cross-Cultural Engagements in Oceania*. London: Routledge, pp.331-59.

Hercus, L., and P. Clarke., 1986. Nine Simpson Desert wells. *Archaeology in Oceania*, 21:51-62.

Hiscock, P., 1985. The need for a taphonomic perspective in stone artefact analysis. *Queensland Archaeological Research*, 2:82-95.

Hiscock, P., 1986. Raw material rationing as an explanation of assemblage differences: A case study from Lawn Hill, Northwest Queensland. In G. Ward (ed.), *Archaeology at ANZAAS*. Canberra: Canberra Archaeological Society, pp.178-90.

Hiscock, P., 1988. A cache of tulas from the Boulia district, Western Queensland. *Archaeology in Oceania*, 23:60-70.

Hiscock, P., 1993. Bondian technology in the Hunter River valley and its implications for the interpretation of late Holocene change in Australia. *Archaeology in Oceania*, 21(1):40-50.

Hiscock, P., 1994. Technological responses to risk in Holocene Australia, *Journal of World Archaeology*, 8(3):267-92.

Hiscock, P., 1996. Mobility and technology in the Kakadu coastal wetlands. *Indo-Pacific Prehistory Association Bulletin*, 15:151-57.

Hiscock, P., 2001. Sizing up prehistory: sample size and composition of artefact assemblages. *Australian Aboriginal Studies*, 1:48-62.

Hiscock, P., and P. J. Hughes., 1980. Backed blades in northern Australia: evidence from northwest Queensland. *Australian Archaeology*, 10:89-95.

Hiscock, P. and S. Mitchell., 1993. *Stone Artefact Quarries and Reduction Sites in Australia: Towards a Type Profile*. Australian Heritage Commission Technical Publication Series No. 4, Canberra: Australian Government Publishing Service.

Hiscock, P., and V. Attenbrow., 1998. Early Holocene backed artefacts from Australia, *Archaeology in Oceania*, 33:49-62.

Hiscock, P. and V. Attenbrow., 2002. Reduction continuums in Eastern Australia: measurement and implications at Capertee 3. In S. Ulm (ed.) *Barriers, Borders, Boundaries*. Tempus volume 7. St Lucia: University of Queensland, pp.167-74.

Hiscock, P., and V. Attenbrow 2003., Early Australian implement variation: a reduction model. *Journal of Archaeological Science*, 30:239-49.

Hiscock, P., and H. Allen., 2000. Assemblage variability in the Willandra Lakes, *Archaeology in Oceania*, 35(3):97-103.

Hiscock, P., and C. Clarkson., 2000. Analysing Australian stone artefacts: an agenda for the twenty first century, *Australian Archaeology*, 50:98-108.

Hodder, I., 1987. The contribution of the long-term. In I. Hodder (ed), *Archaeology as Long-term History*. Cambridge: Cambridge University Press, pp.1-8.

Holdaway, S., 1995a. Stone artefacts and the transition, *Antiquity*, 69:784-97.

Holdaway, S., 1995b. Review of Islands in the Interior. The Dynamics of Prehistoric Adaptations within the Arid Zone of Australia. *Archaeology in Oceania*, 30:43-45.

Holdaway, S., 1995c. Archaeological analysis and the role of statistics: a reply to Veth, *Archaeology in Oceania*, 30:137-38.

Holdaway, S., 2000. Economic approaches to stone artefact raw material variation. In A. Anderson and T. Murray (eds), *Australian Archaeologist: Collected Papers in Honour of Jim Allen*, Canberra: Coombs Academic Publishing, pp.217-30.

Holdaway, S., and N. Porch., 1996. Dates as data. In J. Allen (ed), *Report of the Southern Forests Archaeological Project, Volume 1; Site Descriptions, Stratigraphies and Chronologies*. Bundoora: Archaeology Publications School of Archaeology La Trobe University, pp.251-77.

Holdaway, S., D. Witter, P. Fanning, R. Musgrave, G. Cochrane, T. Doelman, S. Greenwood, D. Pigdon and J. Reeves., 1998. New approaches to open site spatial archaeology in Sturt National Park, New South Wales, Australia. *Archaeology in Oceania*, 33(1):1-19.

Holdaway, S., P.C. Fanning and D.C. Witter., 2000. Prehistoric Aboriginal occupation of the rangelands: interpreting the surface archaeological record of Far Western New South Wales. *Rangelands Journal*, 22(1):44-57.

Holdaway, S. J., P. C. Fanning, M. Jones, J. Shiner, D. C. Witter and G. Nicholls., 2002a. Variability in the chronology of Late Holocene Aboriginal Occupation on the arid margin of Southeastern Australia. *Journal of Archaeological Science*, 29:351-63.

Holdaway. S, P. Fanning, and J. Shiner., 2002b. Geoarchaeological Investigation of Aboriginal Landscape Occupation in Peery National Park, Western New South Wales. Unpublished Report to the New South Wales National Parks and Wildlife Service.

Holdaway, S. P. Fanning and J. Shiner., 2005. 'Absence of Evidence of Evidence of Absence? Understanding the Chronology of Indigenous Occupation of Western New South Wales, Australia'. *Archaeology in Oceania*, 40:33-49.

Holdaway, S., J. Shiner, and P. Fanning., 2004. Hunter-gatherers and the archaeology of the long term: An analysis of surface, stone artefact scatters from Sturt National Park, New South Wales, Australia. *Asian Perspectives*, 43(1):34-72.

Holdaway, S., and N. Stern., 2004. A Record in Stone: The Study of Australia's Flaked Stone Artefacts. Melbourne: Museum Victoria and Aboriginal Studies Press, Australian Institute of Aboriginal and Torres Strait Islander Studies.

Hope, J., 1993. Pleistocene archaeological sites in the central Murray-Darling Basin. In. M. A. Smith, M. Spriggs and B. Fankhauser (eds), *Sahul in Review: Pleistocene Archaeology in Australia, New Guinea and Island Melanesia*. Canberra: Department of Prehistory, Research School of Pacific Studies, The Australian National University, pp.183-96.

Hope, J., A. Dare-Edwards, and M. L. McIntyre., 1983. Middens and megafauna: stratigraphy and dating of the Lake Tandou lunette, western New South Wales. *Archaeology in Oceania*, 18:45-53.

Horton, D. R., 1981. Water and woodland: the peopling of Australia. *AIAS Newsletter*, 16:21-27.

Hughes, P. J., 1977. A Geomorphological Interpretation of Selected Archaeological Sites in Southern Coastal New South Wales. Unpublished PhD thesis, School of Geography, University of New South Wales.

Hughes, P. J., and R. J. Lampert., 1977. Occupation disturbance and types of archaeological deposit, *Journal of Archaeological Science*, 4:135-40.

Hutton, J. T., C. R. Twidale., and A. R. Milnes., 1978. Characteristics and origin of some Australian Silcretes. In T. Langford-Smith (ed), *Silcrete in Australia*. Armidale: Department of Geography, University of New England, pp.19-39.

Jansen, J. D., 2001. Bedrock Channel Morphodynamics and Landscape Evolution in an Arid Zone Gorge: Sandy Creek Gorge, Northern Barrier Range, South-Eastern Central Australia. Unpublished PhD thesis, Macquarie University.

Jelinek, A. R., 1976. Form, function, and style in lithic analysis. In C. E. Cleland (ed), *Cultural Change and Continuity: Essays in Honor of James Bennett Griffin*. New York: Academic Press, pp.19-33.

Johnston, H., 1993. Pleistocene shell middens of the Willandra Lakes. In. M. A. Smith, M. Spriggs and B. Fankhauser (eds), *Sahul in Review: Pleistocene Archaeology in Australia, New Guinea and Island Melanesia*. Canberra: Department of Prehistory, Research School of Pacific Studies, The Australian National University, pp.197-203.

Johnston, H and P. Clark., 1998. Willandra Lakes archaeological investigations 1968-98. *Archaeology in Oceania*, 33:105-119.

Jones, M. and G. Nicholls., 2003. www.Datelab.org.

Jones, R., 1985. *Archaeological Research in Kakadu National Park*. Canberra: Australian National Parks and Wildlife Service.

Jones, R., 1985. Archaeological conclusions. In R. Jones (ed), *Archaeological Research in Kakadu National Park*. Canberra: Australian National Parks and Wildlife Service, pp.291-98.

Jones, R., and I. Johnson., 1985. Deaf Adder Gorge: Lindner Site, Nauwalabila I. In R. Jones (ed), *Archaeological Research in Kakadu National Park*. Canberra: Australian National Parks and Wildlife Service, pp.165-227.

Kamminga, J., 1982. *Over the Edge: Functional Analysis of Australian Stone Tools*. St. Lucia, Brisbane: University of Queensland. Occasional Papers in Anthropology 12.

Kamminga, J., 1985. The pirri graver. *Australian Aboriginal Studies*, 2:2-25.

Kimber, R. G., and M. A. Smith., 1987. An Aranda ceremony. In D. J. Mulvaney and J. P. White (eds), *Australians to 1788*. Broadway: Fairfax, Syme and Weldon Associates, pp.220-38.

Knapp, B. A., and W. Ashmore., 1999. Archaeological landscapes: constructed, conceptualized, ideational. In W. Ashmore and A. B. Knapp (eds), *Archaeologies of Landscape: Contemporary Perspectives*. Oxford: Blackwell, pp.1-30.

Kuhn, S. L., 1991. "Unpacking" reduction: lithic raw material economy in the Mousterian of West-Central Italy. *Journal of Anthropological Archaeology*, 10:76-106.

Kuhn, S. L., 1992. On planning and curated technologies in the Middle Paleolithic. *Journal of Archaeological Research*, 48:185-214.

Kuhn, S. L., 1994. A formal approach to the design and assembly of mobile toolkits. *American Antiquity*, 59(3):426-42.

Kuhn, S. L., 1995. *Mousterian Lithic Technology: An Ecological Perspective*. Princeton: Princeton University Press.

Kvamme, K. L., 1998. Spatial structure in mass debitage scatters. In A. P. Sullivan III (ed.), *Surface Archaeology*. Albuquerque: University of New Mexico Press, pp. 127-41.

Lampert, R. J., and P. J. Hughes., 1988. Early human occupation of the Flinders Ranges. *Records of the South Australian Museum*, 22:139-68.

Langford-Smith, T., 1978. A select review of silcrete research in Australia. In T. Langford-Smith (ed), *Silcrete in Australia*. Armidale: Department of Geography, University of New England, pp.1-11.

Littleton, J., 1998. East and west: burial practices along the Murray River. *Archaeology in Oceania*, 34:1-14.

Littleton, J., 2000. Taphonomic effects of erosion on deliberately buried bodies. *Journal of Anthropological Science*, 27:5-18.

Littleton, J., 2002. Mortuary behaviour on the Hay Plain: do cemeteries exist? *Archaeology in Oceania*, 37:105-122.

Lourandos, H., 1980. Change or stability? Hydraulics, hunter-gathers and population in temperate Australia. *World Archaeology* 11:245-66.

Lourandos, H., 1983. Intensification: a late Pleistocene-Holocene archaeological sequence from southwestern Victoria. *Archaeology in Oceania*, 18:81-94.

Lourandos, H., 1985. Intensification and Australian prehistory. In T. D. Price and J. A. Brown (eds), *Prehistoric Hunter-gathers: the Emergence of Cultural Complexity*. Orlando: Academic Press, pp.385-423.

Lourandos, H., 1996. Change in Australian prehistory: scale, trends and frameworks of interpretation. In S. Ulm, I. Lilley and A. Ross (eds), *Proceedings of the 1995 Australian Archaeological Association Annual Conference, Tempus 6*. St Lucia: Anthropology Museum, The University of Queensland, pp.15-21.

Lourandos, H., 1997, *Continent of Hunter-Gathers: New Perspectives in Australian Prehistory*. Cambridge: Cambridge University Press.

Loy, T., 1993. Prehistoric organic residue analysis: the future meets the past. In M. Spriggs, D. Yen, W. Ambrose, R. Jones and A. Andrews (eds), *A Community of Culture: The People and Prehistory of the Pacific*. Canberra: Department of Prehistory, Research School of Pacific Studies, The Australian National University, Occasional Papers in Prehistory, 21, pp.56-72.

Mabbutt, J. A., 1971. The Australian arid zone as a prehistoric environment. In D. J. Mulvaney. and. J. Golson (eds), *Aboriginal Man and Environment in Australia* Canberra, Australian National University Press. pp.66-79.

Mabbutt, J. A., 1986. Desert Lands. In D. N. Jeans (ed), *Australia: a Geography*. Sydney: Sydney University Press, pp.180-202.

Mardaga-Campbell, M., 1986. Prehistoric living-floors and evidence for them in North Queensland rockshelters. *Australian Archaeology*, 23:42-61.

Martin, S., 1985a. Archaeological Survey of the Popiltah Seismic Survey Lines 85-CP-1, 85-CP-2 and 85-CP-3. Unpublished Report to Claremont Petroleum N. L.

Martin, S., 1985b. Archaeological Survey of the Popiltah Seismic Survey Lines 85-CP-4, 85-CP-5, 85-CP-6, 85-CP-7, 85-CP-8, and 85-CP-9. Unpublished Report to Claremont Petroleum N. L.

Martin, S., and D. O'Donnell., 1995. Stephen's Creek Archaeology Salvage near Broken Hill, Western NSW. In I. Davidson, C. Lovell-Jones and R. Bancroft (eds), *Archaeologists and Aborigines Working Together*. Armidale: University of New England Press, pp.73-75.

McBryde, I., 1984. Kulin greenstone quarries: the social contexts of production and distribution for the Mt William site. *World Archaeology*, 16(2):267-85.

McBryde, I., 1985. Backed blade industries from the Graman rock shelters, New South Wales: some evidence of function. In V. N. Misra and P. Bellwood (eds), *Recent Advances in Indo-Pacific Prehistory: Proceedings of the International Symposium Held at Poona, December 19-21, 1978*. New Delhi: Oxford and IBH Publishing Co, pp.231-49.

McBryde, I., 1997. 'The landscape is a series of stories'. Grindstones, quarries and exchange in Aboriginal Australia: a Lake Eyre case study. In A. Ramos Millán and M. A. Bustillo (eds), *Siliceous rocks and culture*. Granada: Universidad de Granada, pp. 587-607.

McCarthy, F. D., 1949. The prehistoric cultures of Australia. *Oceania*, 19(4):305-19.

McDonald, J., E. Rich and H. Barton. 1994. The Rouse Hill Infrastructure Project (Stage 1) on the Cumberland Plain, western Sydney: recent research and issues. In M. Sullivan, S. Brockwell and A. Webb (eds), *Archaeology in the North*. North Australia Research Unit, Darwin, pp.259-93.

McGlade, J., 1999. The times of history: archaeology, narrative and non-linear causality. In T. Murray (ed), *Time and Archaeology*. London: Rouledge, pp. 139-63.

McGlade, J., and S. E. Van Der Leeuw., 1997. Introduction: archaeology and non-linear dynamics – new approaches to long-term change. In S. E. Van Der Leeuw and J. McGlade (eds), *Time, Process and Structure Transformation in Archaeology*. London: Routledge, pp.1-31.

McIntyre, M. L., 1977. Report of a Survey of Archaeological Sites along the Red Cliffs/ Broken Hill Transmission Lime. Unpublished Report.

McIntyre, M. L., 1981. An archaeological survey of the Mildura-Broken Hill electricity line corridor. In J. Hope (ed), *Darling Surveys 1*, Occasional Papers in Prehistory No. 3, Canberra: Department of Prehistory Research School of Pacific Studies Australian National University, pp.9-32.

McNiven, I. J., 1993. Raw material proximity and bevel-edged tool use, Teewah Beach, Southeast Queensland, *Archaeology in Oceania*, 28:138-43.

McNiven, I. J., 1994. Technological organization and settlement in southwest Tasmania after the glacial maximum. *Antiquity*, 68:75-82.

McNiven, I. J., 2000. Backed into the Pleistocene, *Archaeology in Oceania*, 35(1):48-52.

McPherron, S. and S. Holdaway., 1996. Entrer Trois. In H. Dibble and S. McPherron (eds), A Multimedia Companion to the Middle Paleolithic Site of Combe Capelle Bas (France). CD-ROM. Philadelphia: University of Pennsylvania Museum.

Meltzer, D., 1988. Late Pleistocene human adaptations in eastern North America, Journal of World Prehistory, 2:1-52.

Mills, K., and P. Buckley., 2000. *Koonenberry Excursion Guide*. New South Wales Department of Mineral Resources.

Montet-White, A., 1988. Raw-material economy among medium-sized Late Paleolithic campsites of Central Europe. In H. L. Dibble and A. Montet-White (eds), *Upper Pleistocene Prehistory of Western Eurasia*. Philadelphia: University of Pennsylvania Museum, pp.361-74.

Montet-White, A., 1991. Lithic acquisition, settlements and territory in the Epigravettian of Central Europe. In A. Montet-White and S. Holen (eds), *Raw Material Economies among Prehistoric Hunter-Gathers*. Lawrence: University of Kansas, pp.205-20.

Mulvaney, D. J., 1961. The stone age of Australia. *Proceedings of the Prehistoric Society*, 27:56-107.

Mulvaney, D. J., 1969. *The Prehistory of Australia*. Melbourne: Pelican.

Mulvaney, D. J., 1985. Australian backed blade industries in perspective. In V. N. Misra and P. Bellwood (eds), *Recent Advances in Indo-Pacific Prehistory: Proceedings of the International Symposium Held at Poona, December 19-21, 1978*. New Delhi: Oxford and IBH Publishing Co, pp.211-29.

Murray, T., 1993. Archaeology and the threat of the past: Sir Henry Rider Haggard and the acquisition of time. *World Archaeology*, 25:175-86.

Murray, T., 1999. A return to the 'Pompeii premise'. In T. Murray (ed), *Time and Archaeology*. London: Routledge, pp.8-27.

Myers, F. R., 1986. *Pintupi Country: Pintupi Self: Sentiment, Place and Politics among Western Desert Aboriginies*. Canberra: Australian Institute of Aboriginal Studies.

Nelson, M. C., 1991. The study of technological organization. *Method and Theory in Archaeology*, 3:57-100.

Nelson, M. C., 2000. Abandonment: conceptualization, representation, and social change. In M. B. Schiffer (ed), *Social Theory in Archaeology*. Salt Lake City: University of Utah Press, pp.52-62.

Nelson, M. C., and G. Schachner., 2002. Understanding abandonments in the North American Southwest. *Journal of Archaeological Research*, 10(2):167-206.

Newman, J. R., 1994. The effects of distance on lithic material reduction technology. *Journal of Field Archaeology*, 21:491-501.

New South Wales Department of Mines 1969. 1:250,000 Menindee Geological Map. Geological Series Sheet S1 54-3.

Nicholls, N., and B. Lavery., 1992. Australian rainfall trends during the twentieth century. *International Journal of Climatology*, 12:153-63.

O'Brien, M. J., and T. D. Holland., 1995. Behavioral archaeology and the extended phenotype. In J. Skibo, W. Walker and A. Nielsen (eds), *Expanding Archaeology*. Salt Lake City: University of Utah Press, pp.143-61.

O'Connell, J. F., 1977. Aspects of variation in central Australian lithic assemblages. In R. V. S. Wright (ed), *Stone Tools as Cultural Markers; Change, Evolution, and Complexity*, Canberra: Australian Institute of Aboriginal Studies, pp.269-81.

O'Connell, J. F., 1987. Alyawara site structure and its archaeological implications. *American Antiquity*, 52(1):74-108.

Odell, G. H., 1996. Economizing behavior and the concept of "curation". In G. Odell (ed), Stone Tool: Theoretical Insights into Human Prehistory. New York: Plenum Press, pp.51-80.

Olivier, L., 1999. The Hochdorf 'princely' grave and the question of the nature of archaeological funerary assemblages. In T. Murray (ed), *Time and Archaeology*. London: Routledge, pp.109-38.

Oswalt, W. H., 1976. *An Anthropological Analysis of Food-Getting Technology*. New York: John Wright and Sons.

Page, K., G., Nanson., and D. Price., 1996. Chronology of Murrumbidgee River palaeochannels on the Riverine Plain, southeastern Australia. *Journal of Quaternary Science*, 11:311-26.

Pardoe, C., 1988. The cemetery as a symbol. The distribution of prehistoric Aboriginal burial grounds in southeastern Australia. *Archaeology in Oceania*, 23:1-16.

Pardoe, C., 1991. Competing paradigms and ancient human remains: the state of the discipline. *Archaeology in Oceania* 26:79-84.

Pardoe, C., 1993a. The Pleistocene is still with us: analytical constraints and possibilities for the study of ancient human remains in archaeology. In M. A. Smith, M. Spriggs and B. Fankhauser (eds). *Sahul in Review: Pleistocene Archaeology in Australia, New Guinea and Island Melanesia*. Canberra: Department of Prehistory, Research School of Pacific Studies, The Australian National University. pp.81-94.

Pardoe, C., 1993b. Wamba yadu, a later Holocene cemetery of the central River Murray. *Archaeology in Oceania*, 28:77-84.

Parry, W., and R. Kelly., 1987. Expedient core technology and sedentism. In J. Johnson and C. Morrow (eds), *The Organization of Core Technology*, Boulder: Westview Press, pp.285-308.

Pearce, R. H., 1974. Spatial and temporal distribution of Australian backed blades, *Mankind*, 9(4):300-9.

Pearson, M., and S. Sullivan. 1995. *Looking After Heritage Places: The Basics of Heritage Planning for Managers. Landholders and Administrators*. Melbourne: Melbourne University Press.

Pelcin, A. W., 1997. The effect of core surface morphology on flake attributes: evidence from a controlled experiment. *Journal of Archaeological Science* 24:749-56.

Perlès, C., 1992. In search of lithic strategies: a cognitive approach to prehistoric chipped stone assemblages. In J. C. Gardin and C. Peebles (eds), *Representations in Archaeology*. Bloomington: Indiana University Press, pp.223-47.

Pickard, J., 1994. Post-European changes in creeks of semi-arid rangeland, 'Polpah Station', New South Wales. In A. C. Millington and K. Pye (eds), *Environmental Change in Drylands: Biogeographical and Geomorphological Perspectives*. Chichester: John Wiley and Sons Ltd, pp.271-83.

Plimer, I. R., 1994. *Minerals and Rocks of the Broken Hill, White Cliffs and Tibooburra Districts*. Norwood: Peacock Publications.

Pretty, G.L., 1981. Trial excavation of an Aboriginal open campsite. And site survey, Kinchega National Park. In J. Hope (ed), *Darling Surveys 1*, Occasional Papers in Prehistory No. 3, Canberra: Department of Prehistory Research School of Pacific Studies Australian National University, pp.53-63.

Preucel, R.W., and I. Hodder., 1996. Process, structure and History. In R.W. Preucel and I. Hodder (eds), *Contemporary Archaeology in Theory: A Reader*, Cambridge: Blackwell, pp.205-19.

Raftery, A.E., 1996. Hypothesis testing and model selection. In W.R. Gilks, S. Richardson and D.J. Spiegelhalter (eds), *Markov Chain Monte Carlo in Practice*, London: Chapman and Hall/CRC, pp.163-87.

Rich, E., 1986., Popiltah Seismic Survey Lines 10, 11, 12 Extension, and 1 Extension. Unpublished Report to Claremont Petroleum, N. L.

Richards, J. and N. Ryan., 1985. *Data Processing in Archaeology*. Cambridge: Cambridge University Press.

Richardson, N., 1992. Conjoin sets and strategraphic integrity in a sandstone shelter: Kenniff Cave (Queensland, Australia). *Antiquity*, 66:408-18.

Robins, R., 1984. Cultural resource management in practice: the Moreton Island example. In R. J. Coleman, J. Covacevic, and P. Davie (eds), *Focus on Stradbroke: New Information on North Stradbroke Island and Surrounding Areas, 1974-1984*. Brisbane: Boolarong Publications, pp.33-43.

Robins, R., 1993. Archaeology and the Currawinya Lakes: Towards a prehistory of arid lands of southwest Queensland. Unpublished PhD thesis, Facility of Environmental Sciences, Griffith University.

Robins, R., 1996. A report on archaeological investigations of open hearth sites in southwest Queensland. *Queensland Archaeological Research*, 10:25-35.

Robins, R., 1998. Archaeological investigations at Youlain Springs, Southwest Queensland. *Memoirs of the Queensland Museum, Cultural Heritage Series*, 1(1):57-74.

Robins, R., 1999. Lessons learnt from a taphonomic study of stone artefact movement in an arid environment. In M. J. Mountain and D. Bowdery (eds), *Taphonomy: The Analysis of Processes from Phytoliths to Megafauna*. Canberra: ANH Publications, pp.93-106.

Rognon, P., and M. A. J. Williams., 1977. Late Quaternary climatic changes in Australia and North Africa: a preliminary interpretation. *Palaeogeography, Palaeoclimatology, Palaeoecology*, 21:285-327.

Rolland, N. and H. Dibble., 1990. A new synthesis of Middle Paleolithic variability. *American Antiquity*, 55(3):480-99.

Ross, A., 1981. Holocene environments and prehistoric site patterning in the Victorian Mallee. *Archaeology in Oceania*, 16:145-54.

Ross, A., 1984. If There Were Water: Prehistoric Settlement Patterns in the Victorian Mallee. Unpublished PhD thesis, School of Earth Sciences, Macquarie University.

Ross, A., T. Donnelly and R. Wasson., 1992. The peopling of the arid zone: human-environment interactions. In J. Dodson (ed), *The Naïve Lands*. Melbourne: Longman Cheshire. pp.76-114.

Rossignol, J., 1992. Concepts, methods, and theory building: a landscape approach. In J. Rossignol and L. Wandsnider (eds), *Space, Time and Archaeological Landscapes*, New York: Plenum Press, pp.3-16.

Roth, B., 2000. Obsidian source characterization and hunter-gather mobility: an example from the Tucson Basin. *Journal of Archaeological Science*, 27:305-14.

Roth, B. J., and H. L. Dibble., 1998. Production and transport of blanks and tools at the French Middle Paleolithic site of Combe-Capelle Bas. *American Antiquity*, 63(1):47-62.

Schick, K, 1987. Experimentally derived criteria for assessing hydrologic disturbance of archaeological sites. In D. Nash and M. Petraglia (eds), *Natural Formation Processes and the Archaeological Record*. Oxford: BAR International Series 352. pp.86-107.

Schiffer, M. B., 1972. Archaeological context and systemic context. *American Antiquity*, 37:156-65.

Schiffer, M. B., 1976. *Behavioral Archaeology*. New York: Academic Press.

Schiffer, M. B., 1983. Toward the identification of formation processes. *American Antiquity*, 48(4):675-706.

Schlanger, S. H., 1992. Recognising persistent places in Anasazi settlement systems. In J. Rossignol and L. Wandsnider (eds), Space, *Time and Archaeological Landscapes*. New York: Plenum Press, pp.91-112.

Seeman, M. F., 1994. Intercluster lithic patterning at Nobles Pond: a case for "disembedded" procurement among early Paleoindian societies. *American Antiquity*, 59(2):273-88.

Shanks, M., and C. Tilley., 1987. *Social Theory and Archaeology*. Albuquerque: University of New Mexico.

Shawcross, W., 1998. Archaeological excavations at Mungo. *Archaeology in Oceania*, 33(3):183-200.

Shennan, S., 2000. Population, culture history, and the dynamics of culture change. *Current Anthropology*, 41:811-35.

Sheppard, P., 1997. Characterisation of cherts from sites in southwest Tasmania. *Archaeology in Oceania*, 32:47-53.

Shermer, M., 1995. Exorcising Laplace's demon: chaos and antichaos, history and metahistory. *History and Theory*, 34:59-83.

Shiner, J., 1999. Stone Artefact Assemblage Composition at Stud creek, Sturt national Park, New South Wales, Australia. Unpublished BA (Hons) thesis, Department of Anthropology, The University of Auckland.

Shiner, J, S. Holdaway, H. Allen and P. Fanning. 2005. Understanding Stone Artefact Assemblage Variability in Late Holocene Contexts in Western New South Wales, Australia: Burkes Cave, Stud Creek and Fowlers Gap. In C. Clarkson and L. Lamb (eds), *Rocking the Boat: Recent Australian Approaches to Lithic Reduction, Use and Classification*. British Archaeological Reports International Series. Oxford: BA R Publishing.

Shiner, J., S. Holdaway, H. Allen and P. Fanning., 2007. Burkes Cave and flaked stone assemblage variability in western New South Wales, Australia. *Australian Archaeology*, 64:35-45.

Shott, M. J., 1989. On tool-class use lives and the formation of archaeological assemblages. *American Antiquity*, 54(1):9-30.

Shott, M. J., 1995. How much is a scraper? Curation, use rates, and the formation of scraper assemblages. *Lithic Technology*, 20(1):53-72.

Shott, M. J., 1996a. Innovation and selection in prehistory: a case study from the American Bottom. In G. H. Odell (ed), *Stone Tools: Theoretical Insights into Human Prehistory*. New York: Plenum Press, pp.279-309.

Shott, M. J., 1996b. An exegesis of the curation concept. *Journal of Anthropological Research*, 52:259-80.

Shott, M. J., 1997. Activity and formation as sources of variation in Great Lakes Paleoindian assemblages. *Midcontinental Journal of Archaeology*, 22(2):197-236.

Shott, M. J., 2000. The quantification problem in stone-tool assemblages. *American Antiquity*, 64(4):725-38,

Siiriäinen, A., 1991. Time in archaeology. In T. Hackens., H. Jungner., and C. Carpelar (eds), *Time and Environment: A PACT Seminar, Sept 25-28, 1990, Helsinki, Finland*. Rixensart:Conseil de l'Europe.

Smith, L., 1996. Significance concepts in Australian management archaeology, In L. Smith and A. Clarke (eds), Issues in Management Archaeology. St Lucia: Anthropology Museum, University of Queensland. pp.67-78.

Smith, M. A., 1988. The pattern and timing of prehistoric settlement in Central Australia. Unpublished PhD thesis, University of New England.

Smith, M. A., 1989a. The case for a resident human population in the Central Australian Ranges during full glacial aridity. *Archaeology in Oceania*, 24:93-105.

Smith, M. A., 1989b. Seed gathering in inland Australia: current evidence from seed-grinders on the antiquity of the ethnohistorical pattern of exploitation. In D. R. Harris and G. C. Hillman (eds), *Foraging and Farming: The Evolution of Plant Exploitation*. London: Unwin Hyman, pp.305-17.

Smith, M. A., 1993. Biogeography, human ecology and prehistory in the sandridge deserts. *Australian Archaeology*, 37:35-50.

Smith, M. A., 1996. Prehistory and human ecology in central Australia: an archaeological perspective. In S. R. Morton and D. J. Mulvaney (eds), *Exploring Central Australia: Society, The Environment and the 1894 Horn Expedition*. Chipping Norton: Surrey Beatie and Sones, pp.61-73.

Smith, M. A., and P. M. Clark., 1993. Radiocarbon dates for prehistoric occupation of the Simpson Desert. *Records of the South Australian Museum*, 26(2):121-27.

Smith, M. A., E. Williams, and R. J. Wasson., 1991. The archaeology of the JSN site: some implications for the dynamics of human occupation in the Strzelecki Desert during the Late Pleistocene. *Records of the South Australian Museum*, 25(2):175-92.

Smith, M. E., 1992. Braudel's temporal rhythms and chronology theory in archaeology. In B. A. Knapp (ed), *Archaeology, Annales and Ethno-history*. Cambridge: Cambridge University Press, pp.23-34.

Soil Conservation Service of New South Wales., 1982. 1:250,000 Menindee Land Systems Map. Land Systems Series Sheet S1 54-3. Land Information Centre, New South Wales Department of Lands.

Soil Conservation Service of New South Wales ,. 1991. Land Systems of Western New South Wales. Technical Report No. 25, Soil Conservation Service of New South Wales.

Stern, N., 1993. The structure of the Lower Pleistocene archaeological record: A case study from the Koobi Fora Formation. *Current Anthropology*, 34(3):201-25.

Stern, N., 1994. The implications of time-averaging for reconstructing the land-use patterns of early tool-using Hominids. *Journal of Human Evolution*, 27:89-105.

Stern, N., 1995. The "blue" tuff locality at Koobi Fora in northern Kenya: archaeological time and the record of past human behaviour. *The Artefact*, 18:49-59.

Stern, N., N. Porch., and I. McDougall., 2002. Fxjj43: A window into a 1.5-million-year-old Palaeolandscape in the Okote member of the Koobi For a formation, Northern Kenya. *Geoarchaeology*, 17(4):349-92.

Stevens, B. P. J., 1986. Post-depositional history of the Willyama Supergroup in the Broken Hill Block, NSW. *Australian Journal of Earth Sciences*, 33:73-98.

Stevenson, M. C., 1982. Towards an understanding of site abandonment behavior: evidence from historic mining camps in the southwest Yukon. *Journal of Anthropological Archaeology*, 1:237-65.

Stockton, E, D., 1973. Shaw's Creek shelter: human displacement of artefacts and its significance. *Mankind*, 9:112-17.

Sturt, C., 1849. Narrative of an Expedition into Central Australia. London: T and W Boone.

Sullivan A. P. III., 1992. Investigating the archaeological consequences of short-duration occupations. *American Antiquity*, 57(1):99-115.

Sullivan A. P. III., 1998. Surface phenomena in archaeological research. In A. P. Sullivan III (ed.), *Surface Archaeology*. Albuquerque: University of New Mexico Press, pp. XI-XIII.

Sullivan, M. E. and S. Simmons., 1979. Silcrete: a classification for flaked stone artefact assemblages. *The Artefact*, 4:51-60.

Taçon, P. S. C., 1991. The power of stone: symbolic aspects of stone use and tool development in western Arnhem Land, Australia. *Antiquity*, 65:192-207.

Taçon, P. S. C., 1994. Socialising landscapes: the long-term implications of signs, symbols and marks on the land. *Archaeology in Oceania*, 29:117-29.

Theunissen, R., J. Balme and W. Beck., 1998. Headroom and human trampling: cave ceiling-height determines the spatial patterning of stone artefacts at Petzkes Cave, northern New South Wales. *Antiquity*, 72:80-89.

Thomas, D. H., 1975. Nonsite sampling in archaeology: up the creek without a site? In J. W. Mueller (ed), Sampling in Archaeology. Arizona: The University of Arizona Press. pp.61-81.

Thorley, P. B., 1998a. Pleistocene settlement in the Australian arid zone: occupation of an inland riverine landscape in the Central Australian Ranges. *Antiquity*, 72:34-45.

Thorley, P. B., 1998b. Shifting Location, Shifting Scale: A Regional Landscape Approach to the Prehistoric Archaeology of the Palmer River Catchment, Central Australia. Unpublished PhD thesis, Northern Territory University.

Thorley, P. B., 1999. Regional archaeological research in the Palmer River catchment. *Australian Aboriginal Studies*, 2:62-68.

Thorley, P. B., 2001. Uncertain supplies: water availability and regional archaeological structure in the Palmer River Catchment, central Australia. *Archaeology in Oceania*, 36:1-14.

Thorley, P., L. Warren., and P. Veth., 1994. Wardaman point technology: an examination of sites recorded during CRM investigations in the Victoria River region. In M. Sullivan, S. Brockwell and A. Webb (eds), *Archaeology in the North*. North Australia Research Unit, Darwin. pp.57-71.

Thorne, A. G., 1976. Morpological contrasts in Pleistocene Australians. In R. L. Kirk and A. G. Thorne (eds), *The Origins of the Australians*. Canberra: Australian Institute of Aboriginal Studies, pp.95-112.

Thorne, A.G., 1977. Separation or reconciliation? biological clues to the development of Australian society. In J. Allen, J. Golson and R. Jones (eds), *Sunda and Sahul*. London: Academic Press, pp.187-204.

Thorne, A., R. Grun, G. Mortimer, N. A. Spooner, J. J. Simpson, M. McCulloch, L. Taylor, and D. Curnoe., 1999. Australia's oldest human remains: age of the Lake Mungo 3 skeleton. *Journal of Human Evolution*, 36:591-612.

Tindale, N. B., 1941. The antiquity of Man in Austrlia. The Austrlian Journal of Science:144-47.

Tomka, S. A. and M. G. Stevenson, 1993. Understanding abandonment processes: Summary and remaining concerns, In C. M. Cameron and S. A. Tomka (eds), *Abandonment of Settlements and Regions: Ethnoarchaeological and Archaeological Approaches*. Cambridge: Cambridge University Press, pp.191 -95.

Torrence, R., 1989. Retooling: towards a behavioural theory of stone tools. In R. Torrence (ed.), *Time, Energy and Stone Tools*. Cambridge: Cambridge University Press, pp.57-66.

Trigger, B. G., 1991. Distinguished lecture in archaeology: constraint and freedom-a new synthesis for archaeological explanation. *American Anthropologist*, 93:551-69.

Veth, P.M., 1989. Islands in the interior: a model for the colonization of Australia'a arid zone. *Archaeology in Oceania*, 24:81-92.

Veth, P.M., 1993. *Islands in the Interior: The Dynamics of Prehistoric Adaptations within the Arid Zone of Australia*. Ann Arbor: International Monographs in Prehistory.

Veth, P.M., 1996. Current archaeological evidence from the Little and Great Sandy Deserts. In P. Veth and P. Hiscock (eds), *Archaeological of Northern Australia: Regional Perspectives*. Tempus Volume 4, St Lucia: Anthropology Museum, The University of Queensland, pp.50-65.

Veth, P., G. Hamm, and R. J. Lampert., 1990. The archaeological significance of the lower Cooper Creek. *Records of the South Australian Museum*, 24(1):43-66.

Villa, P., 1982. Conjoinable pieces and site formation processes. *American Antiquity*, 47:276-90.

Vinnicombe, P., 1984. Single sites or site complexes? A case study from north of the Hawkesbury, New South Wales. In S. Sullivan and S. Bowdler (eds), *Site Surveys and Significance Assessment in Australian Archaeology*. Canberra: Department of Prehistory, Research School of Pacific Studies, Australian National University, pp.107-18.

Vita-Finzi, C. and E. S. Higgs., 1970 Prehistoric economy in the Mount Carmel area of Palestine: site catchment analysis. *Proceedings of the Prehistoric Society*, 36:1-37.

Wandsnider, L., 1992. The spatial dimension of time. In J. Rossignol and L. Wandsnider (eds), *Space, Time and Archaeological Landscapes*, New York: Plenum Press, pp.257-81.

Wandsnider, L., 1996. Describing and comparing archaeological spatial structures. *Journal of Archaeological Method and Theory*, 3:319-84.

Wandsnider, L., 1998a. Landscape element configuration, lifespace, and occupation history: ethnoarchaeological observations and archaeological applications. In A. P. Sullivan III (ed.), *Surface Archaeology*. Albuquerque: University of New Mexico Press, pp.21-39.

Wandsnider, L., 1998b. Regional scale processes and archaeological landscape units. In A. Ramenofsky and A. Steffen (eds), *Unit Issues in Archaeology: Measuring Time, Space and Material*. Salt Lake City: University of Utah Press, pp.87-102.

Wandsnider, L. and J. I. Ebert., 1988. Issues in archaeological surface survey: meshing method and theory. *American Archaeology*, 7(1):2

Warner, R. F., 1986. Hydrology. In D. N. Jeans (ed), *Australia: a Geography*. Sydney: Sydney University Press. pp.49-79.

Wasson, R. J., 1979. Sedimentation history of the Mundi Mundi alluvial fans, Western New South Wales. *Sedimentary Geology*, 22:21-51.

Wasson, R. J., and R. W. Galloway., 1986. Sediment yield in the Barrier Range before and after European settlement. *Rangelands Journal*, 8(2):79-90.

Waters, M. R., 1992. *Principles of Geoarchaeology: A North American Perspective*. Tucson: University of Arizona Press.

Waters, M. R., and D. D. Kuehn., 1996. The geoarchaeology of place: the effect of geological processes on the preservation and interpretation of the archaeological record. *American Antiquity*, 61(3):483-97.

Webb, C., 1993. The lithication of a sandy environment. *Archaeology in Oceania*, 28:105-11.

Webb, C., 1996. The formation of a surface archaeological record in semi-arid Australia. Unpublished PhD thesis, School of Archaeology, La Trobe University.

Webb, S. G., 1989. *The Willandra Lakes Hominids*. Canberra: Department of Prehistory, Research School of Pacific Studies Australian National University.

Wheatley, D. and M. Gillings., 2002. *Spatial Technology and Archaeology*. London: Taylor and Francis.

Willey, G. R., 1953. *Prehistoric Settlement Patterns in the Virú Valley, Peru*. Washington: Bureau of American Ethnology, Bulletin no. 155.

Williams, E., 1988. The archaeology of the Cooper Basin: report on fieldwork. *Records of the South Australian Museum*, 22(1):53-62.

Williams, E., 1998. The archaeology of lake systems in the middle Cooper Basin, North-eastern South Australia. *Records of the South Australian Museum*, 30(2):69-91.

Williams, M. A. J., D. L Dunkerley., P. De Decker., A. P. Kershaw and T. Stokes., 1993. *Quaternary Environments*. London: Edward Arnold.

Witter, D., 1989. Archaeological Investigations of a Water Reservoir on "Thackaringa". Unpublished Report, New South Wales National Parks and Wildlife Service.

Witter, D., 1992. Regions and Resources. Unpublished PhD thesis, Department of Prehistory, Australian National University.

Wobst, H. M., 1978. The archaeo-ethnology of hunter-gathers of the tyranny of the ethnographic record in archaeology. *American Antiquity* 43(2):303-309.

Wobst, H. M., 1983. We can't see the forest for the trees: sampling and the shapes of archaeological distributions. In J. Moore and A. S. Keene (eds), *Archaeological Hammers and Theories*. New York: Academic Press, pp.38-85.

Yellen, J. E., 1977. *Archaeological Approaches to the Present: Models for Reconstructing the Past*. New York: Academic Press.

Yoffee, N., and A. Sherratt., 1993. Introduction: the sources of archaeological theory. In N. Yoffee and A. Sherratt (eds), *Archaeological Theory: Who Sets the Agenda?*, Cambridge: Cambridge University Press, pp.1-9.

Zvelebil, M., S. W. Green and M. G. Macklin., 1992. Archaeological landscapes, lithic scatters, and human behavior. In J. Rossignol and L. Wandsnider (eds), *Space, Time and Archaeological Landscapes*, New York: Plenum Press, pp.193-226.

www.ingramcontent.com/pod-product-compliance
Lightning Source LLC
LaVergne TN
LVHW071218210926
848482LV00015B/838
* 9 7 8 1 4 0 7 3 0 2 5 2 2 *